ADVANCES IN PHYSICAL METALLURGY

In accordance with the Bye-Laws
under the ROYAL CHARTER
and by Special Resolution of the Council
at its meeting on the third day of July 1989
SIR ALAN COTTRELL Kt.
was elected an HONORARY FELLOW of
The Institute of Metals
and in confirmation
the Seal of the Institute was affixed on
the twenty-seventh day of September 1989
President
Secretary

ADVANCES IN
PHYSICAL METALLURGY

A collection of invited papers presented to mark the 70th birthday year of Professor Sir Alan Cottrell

Edited by
J. A. Charles and G. C. Smith

The Institute of Metals

CRC Press
Taylor & Francis Group
Boca Raton London New York

CRC Press is an imprint of the
Taylor & Francis Group, an **informa** business

First published 1990 by The Institute of Metals

Published 2021 by CRC Press
Taylor & Francis Group
6000 Broken Sound Parkway NW, Suite 300
Boca Raton, FL 33487-2742

ISBN 13: 978-0-901462-85-5 (hbk)

British Library Cataloguing in Publication Data

Advances in physical metallurgy.
1. Physical metallurgy
I. Charles, James A. (James Anthony) *1926-* II. Smith G. C.
III. Cottrell, *Sir,* Alan *1919-*
669.9

Library of Congress Cataloging-in-Publication Data

Typeset by
Fakenham Photosetting Ltd,
Fakenham, Norfolk

Contents

Preface

Sir Alan Cottrell was 70 on 17 July 1989. The significance of the date had not escaped the notice of Professor Robert Cahn who pointed it out to us with the suggestion that there might be an occasion during the year to celebrate both the anniversary and the distinction of Sir Alan.

It was eventually decided that the celebration would take the form of a one-day meeting and dinner in Cambridge, broadly surveying the position in the areas of physical metallurgy in which he had been, and still is, so powerful a force, hoping particularly that colleagues and friends who had worked with him would be able to attend for an informative but not too serious a gathering.

The response to initial enquiries was wholly encouraging and arrangements went ahead for the meeting on 29 September, with ~100 registered delegates attending, some from overseas, and many more coming for the lectures themselves from the Cambridge area.

The papers presented were both nostalgic in their surveys of the contributions by Sir Alan in so many fields, and progressive in indicating how it had been possible to develop from the sure foundations that he had laid. Sir Alan himself showed all the old flair and capability for originality in his own paper on the strength of grain boundaries, an aspect of his present interests which has led to recent papers in The Institute of Metals Journal of Materials Science and Technology.

In view of the number of papers presented, the time available for discussion was limited. However the contributions which were made revealed further aspects of Sir Alan's influence on the development of physical metallurgy.

His name will always be known to succeeding generations of materials scientists and in this context we were delighted that contributions from industrial companies towards the cost of the meeting have also enabled us to create a Sir Alan Cottrell prize in the Department of Materials Science and Metallurgy at Cambridge. This will be a permanent reminder of his associations with the Department, and will be awarded annually to an outstanding undergraduate.

On such a birthday gifts were appropriate, and as well as personal mementos presented at the dinner from the delegates and from some unable to attend personally, the occasion was marked by the presentation of an Honorary Fellowship of The Institute of Metals by its Secretary, Dr Ashley Catterall.

The Institute of Metals was, in fact, closely associated with the meeting and most generously undertook to cover the initial cost of the publication of these proceedings, and has also arranged for Sir Alan to receive a leatherbound presentation copy of the book. The success of the meeting owed so much to the cooperation of those approached to present papers and to those who took part in the discussions. In the background, also, much depended on the good offices of Mrs P. Summerfield, who acted as Secretary to the Conference, Mr De Castro at the controls in the Babbage Lecture Theatre, and Mr W. Pickup who looked after the production of the book. We thank you all.

April 1990

J. A. Charles and G. C. Smith
Department of Materials Science and Metallurgy, Cambridge

Acknowledgements

Contributions from the following, to the Sir Alan Cottrell prize, are gratefully acknowledged:

British Petroleum International Ltd
British Steel plc
Cookson Group plc
General Electric Company – Research
Imperial Metal Industries Titanium Ltd
National Power
Pilkington plc
Rolls Royce
STC Technology Ltd
United Kingdom Atomic Energy Authority
Welding Institute

Welcoming Comments

D. Hull

Goldsmiths' Professor of Metallurgy, University of Cambridge

It is a pleasure and a privilege for me to welcome you all to Cambridge for this, all too brief, celebration of the 70th birthday year of Sir Alan Cottrell. It is particularly appropriate that this meeting is concerned with 'Advances in Physical Metallurgy'. Alan Cottrell's contributions to the development of the subject have been outstanding. It is interesting to speculate on where the subject would have been now without his contributions. For instance, I know from personal experience that many scientists have been drawn to the subject by the inspiration he has provided.

This meeting has been organised by Jim Charles and Gerry Smith after Robert Cahn had sown the seeds of an idea. I think they are to be congratulated on the success they have achieved in bringing together so many of 'the great and the good' in physical metallurgy. The programme of today's meeting reflects some of the major areas in physical metallurgy in which Alan Cottrell has worked and, in some cases, is still working. The papers are given by colleagues who have worked closely with him, and by scientists who have been guided and inspired by him, in their formative years. Their willingness to contribute and their distinction in their field is a tribute in itself to Alan Cottrell's work. The speakers come from all periods of Sir Alan's career: a Lecturer and Professor of physical metallurgy at Birmingham University, a Deputy Director of the Metallurgy Division at Harwell, Goldsmiths' Professor at Cambridge University and Master of Jesus College here at Cambridge. His influence spreads over many decades and is still with us. I was delighted when he accepted my invitation to join the Department again after he completed his term as Master of Jesus College. He is a keen and active member of the Department. I am sure you are all familiar with his highly acclaimed new book *Introduction to the Modern Theory of Metals*, much of which he wrote in the Department. His work is also valued by students who benefit from a series of lectures on the metallic state, and by many research scientists who have ready access to him in the coffee room. It won't have gone unnoticed that he is also contributing an original paper to this meeting. I suspect that new ideas on fracture at grain boundaries are few and far between these days, so we have a treat in store.

In passing I should also mention the other members of, what I lovingly like to call, the Department's 'geriatric wing' who make so many valuable contributions: Gerry Smith, Robert Cahn, Elizabeth Yoffe, Jack Mayne and of course, AHC's distinguished successor in the Goldsmiths' Chair, Robert Honeycombe. I have noted that there are two ex-Goldsmiths' Professors and one present Goldsmiths' Professor here today. What may not be so obvious is that there are also two ex-Henry Bell Wortley Professors and one present Henry Bell Wortley Professor. This, of course, is part of the Liverpool story. Walter Owen and I are the ex-Professors and Colin Humphreys is the present Henry Bell Wortley Professor. This gives me an opportunity to say two things. Firstly, in Walter Owen's name, welcome to those who have come from abroad for this celebration — we are delighted to see you all. Secondly, greetings and best wishes to Colin Humphreys who has accepted Cambridge University's invitation to the new Chair of Materials Science in the Department. He will take up his appointment in January 1990.

I would like to thank the Institute of Metals who have so generously agreed to publish the proceedings of this meeting and to give a copy of the book to all the participants. At the Celebration Dinner tonight the Secretary General of the Institute, Dr Ashley Catterall, will be presenting Sir Alan with the Honorary Fellowship of the Institute. Yet another tribute.

May I end on a reflective note. When I was being interviewed in 1964 for the new Chair of Physical Metallurgy at Liverpool, my CV was full of dislocations. I was asked by Professor Jim Cassells, a very formidable nuclear physicist, what I was going to do for my research in a couple of years time when all the dislocation problems had been solved. 'What will you do for the rest of your 32 years as a Professor?' Naturally, I vigorously defended 'dislocations' and although I now blush to say it, suggested that there were enough challenges in dislocation concepts to keep me going for the rest of my career. It has not been so, but I do believe that the underlying concepts of physical metallurgy, of which dislocations are a particularly good example, are as important today as they were 25 years ago and they have had, and can have, a profound influence on the way we think about all classes of materials.

I look forward with real interest to the perspectives our speakers give today to 'Advances in Physical Metallurgy' as we sit at the feet of our guru — Alan Cottrell.

1 Dislocations and Cracks

B. A. Bilby

Emeritus Professor, Department of Mechanical and Process Engineering, University of Sheffield

Introduction

This is not the place to attempt a comprehensive survey of Alan Cottrell's contributions to the theory of dislocations and cracks. Instead, I shall be selective and try to give some flavour of the times when the work was done.

I first met Alan when I came from the Admiralty to Birmingham as a research assistant in 1946. Metallurgy was housed at Edgbaston in one of the old high blocks radiating from the ring of buildings centred on 'Joe', the vast tower erected in honour of Joseph Chamberlain. There was a central corridor open to a lofty roof, with offices to the right and laboratories to the left; and on top of the offices and labs, large galleries on each side, where orchestras might have played. Before I left in 1951 the corridor was to be roofed over — to the accompaniment of much pneumatic drilling — and the galleries replaced by more offices.

In 1946 Alan had a small lofty room on the right-hand side separated from the main corridor by an anteroom which contained one of the few telephones of the department. In the lab opposite Syd Harper had built a small Andrade tensile apparatus with a beam and mirror and was pulling single crystals. Alan was supervising the construction of another apparatus with cams for single crystal constant stress creep experiments, later to be carried out with Aytekin. At the inner end of the galleries a short stairway climbed yet higher to 'The Dome' — a vast circular hall which led, if one turned left, to the uppermost of the great circumferential corridors connecting the various radiating departments of the University. Here in the dome, among others, worked Jaswon and Wheeler on martensite.

The times were exciting. People everywhere were returning to more fundamental research, with high hopes that science could help with the peace as it had with the war. The next ten years were to see remarkable advances in the field in which Alan had chosen to work. The Italian mathematicians and others had worked on the dislocation as a many valued solution to the elastic equations, but the literature on dislocations in crystal plasticity was sparse and there was no direct evidence for them. 'A dislocation', Hume Rothery wrote to me in jest in 1950, 'is an invention of a mathematician who cannot understand the properties of metals.' The theory of the strength and flow of metals and alloys was a neglected subject, despite the considerable experimental effort expended by a number of pioneers on fundamental studies and the vast labour daily undertaken during the development of industrial alloys. The time was thus ripe for dislocation theory to be taken seriously. Studies expanded at a number of centres throughout the world, and those carried out in the United Kingdom at Birmingham, Bristol and Cambridge played a leading role in the developments which ensued.

Early Work on Dislocations

The importance of the flexibility of dislocations in influencing their behaviour in solid solutions and precipitation hardened alloys had been emphasised by Mott and Nabarro.[1] After reading their paper, Alan remarked that he thought that solute atoms too could move under the interaction. This led to his theory of the migration of solute atoms in the dislocation field,[2] which spawned much work on the yield point, strain ageing and blue brittleness and the influence of dislocations and other sources of internal stress on precipitation and the distribution of point defects.

Alan's quantitative and effective application of his idea quickly led for the first time to a wide recognition of its importance, though the notion of solute atom migration to relieve stress,[3] even to a dislocation,[4] had been aired before. His manner of establishing the interaction formula is an early example of an ability he was to demonstrate again and again. He announced at the time, when there was much debate among the research workers concerned about the theory of these interactions, that one should simply calculate the work done against the stress field when the displacements associated with the replacement of a solvent atom by a solute atom take place, a result which was soon discussed and justified by a number of authors using the formal theorems of elasticity theory.[5,6,7] Norman Petch is to speak next about the yield point, so I will say no more about this work, except to note the very considerable interest it aroused and the many papers from Birmingham which it inspired.

By 1954, the year after Alan's book[8] on Dislocations and Plastic Flow in Crystals appeared, some fifteen papers had been published on this topic. They dealt with various aspects of the theory of the elastic, chemical, electrical and geometrical interactions of a solute atom and a dislocation when one or the other or both are moving,[9–11] particularly including the $t^{2/3}$ law in strain ageing,[12] the migration to moving dislocations[13] and the electrical interaction;[14] and also many experimental observations: on yield points or quench and strain ageing (including resistivity changes) due to carbon and nitrogen in iron,[15–19] nitrogen in single crystals of cadmium,[19] zinc[20] and β-brass;[21] and serrated yielding.[21, 22] During this period, the theory of dislocations advanced rapidly and evidence for their existence multiplied. A striking result here was the confirmation of the dislocation as a preferential site for precipitation,[1] as shown by the decoration techniques.

Alan's contributions to dislocation theory in this period were not limited to those associated with solute atom interactions. In a long article[23] he again deduces important effects by simple arguments: for example, consideration of the bending produced by edge dislocations in finite crystals is made to provide a theory of polygonisation; and the principle of virtual work is used to deduce the magnified stress at a dislocation pile-up.

Other interests of this period are also to be discussed here later. They include the important note on the formation of Cottrell–Lomer locks[24] and papers on dislocations in twinning[25] and in ordered structures.[26] He also interested himself in creep,[27–30] including a study of surface effects[19, 31] and in the role of jogs and the 'forest' of dislocations intersecting a plane, acting as obstacles to slip on it. He discussed in 1952[32] the interaction energy for the cutting of these obstacles. There followed a number of theoretical and experimental papers associated with hardening, recovery and the migration of point defects.[33–37]

From the Harwell period came the paper on creep due to irradiation growth,[38] a note on the formation of vacancies and interstitials,[39] and further studies of quench, work, and irradiation hardening.[40–43]

Dislocations and Cracks

His active interest in fracture also began about this time. I recall that in one of our early discussions at Birmingham, Alan remarked to me that 'fracture was a very messy subject'; and so it was, when one compared it with the beautiful order which dislocation theory was bringing to the description of crystal plasticity on the microscale when few dislocations were involved (fortunately, in those early days we could not see directly the great complexities associated with dislocations en masse, or perhaps our enthusiasm might have been a little damped!).

Characteristically, Alan was soon to change radically the situation in fracture and to make major contributions to our understanding in this field also. The first incursion was a discussion about possible dislocation processes which might cause the intrusions and extrusions observed in fatigue.[44] This was soon followed by important papers on brittle fracture in steels.[45, 46] Arguing from the effect on fracture of hydrostatic tension, it is emphasised that the controlling process in the transition to macroscopic failure must be the growth of a microcrack, rather than its formation from inclusions or precipitates or by twinning or slip. The same conclusion was reached independently at this time by Petch,[47] through his work on the friction stress. A possible mechanism for wedge crack nucleation by double slip on inclined planes was suggested,[46] which would provide for easy nucleation even in the absence of inclusions and avoid the nucleation control implied by the simple pile-up mechanism.[48] Combining this wedge crack mechanism with the theory of the yield point in polycrystals leads to a detailed theory of the ductile–brittle transition and enables various predictions to be made about the behaviour above and below it, although it does not include the effect of carbide size. This paper[46] also discusses the distinction between what Alan later called cumulative and non-cumulative fractures,[49] and the mechanics of catastrophic fully ductile fracture in shells and plates.

It was perhaps a short step from the balance between slip dislocations and wedge cracks to describe the crack itself by a dislocation array,[50] a step taken also independently in 1959 by Friedel[51] in a paper at the same meeting. Here, too, Alan set out for the first time his views on ductile fracture as a process of internal necking due to plastic instabilities associated with the exhaustion of hardening capacity rather than the exhaustion of ductility, with inclusions or precipitates as the major source of cavity nuclei.

Further understanding of ductile fracture became possible when, in 1960, he used[52] the estimate of the critical crack tip opening displacement u_f given by a very simple model of plastic relaxation at the crack tip based on dislocation theory,[53] to discuss various types of fracture, noting that $\sigma_y u_f$ was essentially the effective surface energy of fracture. A similar model of the plasticity, but without the use of dislocations or crack dislocations, and under tension, was given also in 1960 independently by Dugdale.[54] This was used to discuss plastic zones round slits in steel sheets, but not to obtain the crack tip displacement. The use of the crack tip displacement in fracture was discussed independently by Wells.[55, 56] It is interesting to note that in his original paper,[55] the term COD meant, for him, crack opening *dislocation* and not the crack opening displacement now widely associated with the aeronym. In this paper,[55] Wells used a different approximate method to estimate u_f, relating it to the crack extension force and stress intensity factor in small scale yielding.

Clearly, I cannot attempt here any comprehensive account of the subsequent exploitation of these DBCS or strip yield models, which continues to this day. (For a brief review up to 1968, see Ref. 57; related procedures for removing the crack tip singularity have been used by Barenblatt[58, 59] and Vitvitskii and Leonov.[60]) Alan himself made use of this model in further work on fracture[61–64] and particularly in the Tewkesbury lecture,[49] in the discussion of notch sensitivity and the classification of various types of fracture. He also made other specific studies of the effect of notches on fracture.[65, 66] It is interesting to note that during this period of great activity in the fracture field he still found time to produce another book.[67] Later, after turning his attention to the ultimate strength of solids[68, 69] and the balance between ductility and brittleness[70, 71] he gave a novel treatment[72] of a long studied problem, namely the crack with a force between its faces more general than is implied by the simple DBCS model.[49, 63] The novelty, which again is characteristic of his direct approach was to obtain general expressions for the force and displacement across the cracking surface by taking a distribution of simple Griffith cracks from lengths c_1 to c_2; this ensures automatically that the force vanishes when $x < c_1$ and that the displacement vanishes when $x > c_2$. Of course, as with any inverse method for the solution of the integral equation which really governs the problem,[57] one must take care that the force law deduced is independent of crack length.

The crack tip displacement derived from the DBCS model played an important part in a development which has now a wide practical application. Alan himself,[49, 63] using a critical displacement criterion, showed how the fracture strength could fall from the stress σ_l in the plastic zone to a small value characteristic of Griffith (low stress) fracture, as the parameter $c\epsilon_l/b_l$ increased. (Here c = crack length, ϵ_l = elastic strain corresponding to σ_l and b_l = crack tip opening displacement.) For a mode of fracture characterised by σ_l and b_l he defined $c \simeq b_l/\epsilon_l$ as the crack length at which notch sensitivity began. With $\sigma_l = \sigma_y$ and a constant crack tip displacement b_l he was then able to explain fractures below general yield in large structures.[49, 63] Heald, Spink and Worthington[73] took σ_l to be σ_u, the UTS (partly on the grounds that the failure stress $\sigma_f = \sigma_u$ when $c = 0$). They were then able to correlate results from small specimens, 'invalid' by ASTM standards, with the behaviour of large specimens, again using the DBCS interpolation line with constant crack opening displacement. Next, Dowling and Townley[74] introduced a diagram where σ_f/σ_l is replaced by L_f/L_u, where L_u is the plastic collapse load and L_f the failure load; while $c\epsilon_l/b_l$ is replaced by L_K/L_u, where L_K is the LEFM failure load. The DBCS line is thus now an interpolation between failure by LEFM and failure by plastic collapse. Finally, Harrison, Loosemore and Milne[75] interchanged the axes, representing LEFM by the ordinate K/K_{IC}, with the abscissa L/L_u

now the applied load over the plastic collapse load; this produced the original R6 failure assessment diagram, with the failure assessment line again an interpolation based on the infinite medium DBCS curve. Since the DBCS model has no wake, caution is needed in using such a diagram to discuss crack growth,[76] although procedures for doing this were at one time advocated, see, for example.[77–79] In the comprehensive R6 assessment procedure subsequently developed[80, 81] the DBCS interpolation has now been abandoned and the discussion of crack growth in the most sophisticated procedure (Option 3, Category 3) is stated to correspond to analysis by J_R curves, although it is not yet clear how complete this correspondence is.

It is natural to attempt to extend the very simple DBCS model of plastic relaxation at the crack tip. The use of double slip lines[82–84] and of superdislocations[85, 86] has been developed by a number of authors. Representation of the plastic field by a continuous distribution of dislocations has also received some consideration and has been used, for example, to confirm[86, 87] explicitly for small scale yielding that there is no crack extension force on the relaxed edge crack in autoplane strain considered by Hult and McClintock.[88]

Conclusion

We have tried in this brief account of some of Alan's work on dislocations and cracks to indicate how widely he has stimulated others by his simple direct approach. But his contributions to these two topics form only a part of the great influence he has had over a wider scientific and technological field, as we are to hear in the rest of this meeting.

References

1. N. F. Mott and F. R. N. Nabarro, in *Report of a Conference on Strength of Solids*, Bristol, 1–19, July 1947. The Physical Society, London, 1948.
2. A. H. Cottrell, *ibid*, pp. 30–38.
3. W. S. Gorsky, *Phys. Z. Sowjet.* **8**, 457, 1935.
4. H. Sutton and J. W. Wilstrop, *J. Inst. Metals* **60**, 277, 1937. (Discussion to M. L. Gayler, *Ibid*), 249.
5. G. Leibfried, *Z. Phys.* **126**, 781, 1949.
6. B. A. Bilby, *Proc. Phys. Soc., Lond.* **A61**, 191, 1950.
7. J. D. Eshelby, *Phil. Trans. Roy. Soc.* **A244**, 87, 1951.
8. A. H. Cottrell, 'Dislocations and Plastic Flow in Crystals'. Clarendon Press, Oxford, 1953.
9. A. H. Cottrell, in *A Symposium on the Plastic Deformation of Crystalline Solids*,

Pittsburgh, May 1950, pp. 60–76; Carnegie Institute of Technology and ONR, NAVEXOS-P-834, 1950.

10. A. H. Cottrell, in *L'Etat Solide*, Neuvième Conseil de Physique, Bruxelles, Septembre 1951, pp. 487–502; Institut International de Physique Solvay, Stoops, Bruxelles, 1952.
11. A. H. Cottrell, in *Relation of Properties to Microstructure*, Seminar at the 35th National Metal Congress and Exposition, Cleveland, Ohio, October 1953, pp. 17–23; Cleveland, Ohio, ASM, 1954.
12. A. H. Cottrell and B. A. Bilby, *Proc. Phys. Soc., Lond.* **A62**, 49, 1949.
13. A. H. Cottrell and M. A. Jaswon, *Proc. Roy. Soc.* **A199**, 104, 1949.
14. A. H. Cottrell, S. C. Hunter and F. R. N. Nabarro, *Phil. Mag.* **44**, 1064–1067, 1953.
15. A. H. Cottrell and A. T. Churchman, *J. Iron and Steel Inst.* **162**, 271–276, 1949.
16. A. H. Cottrell and A. T. Churchman, *Nature* **167**, 943, 1951.
17. A. H. Cottrell and G. M. Leak, *J. Iron and Steel Inst.* **172**, 301–306, 1952.
18. S. Harper, *Phys. Rev.* **83**, 709–712, 1951.
19. A. H. Cottrell and D. F. Gibbons, *Nature,* **162**, 488, 1948.
20. H. L. Wain and A. H. Cottrell, *Proc. Phys. Soc., Lond.* **B63**, 339, 1950.
21. G. W. Ardley and A. H. Cottrell, *Proc. Roy. Soc. Lond.* **A219**, 328–341, 1953.
22. A. H. Cottrell, *Phil. Mag.* **44**, 829–832, 1953.
23. A. H. Cottrell, *Progr. Met. Phys.* **1**, 77, 1949.
24. A. H. Cottrell, *Phil. Mag.* **43**, 645–647, 1952.
25. A. H. Cottrell and B. A. Bilby, *Phil. Mag.* **42**, 573–581, 1951.
26. W. A. Rachinger and A. H. Cottrell, *Acta Met.* **4**, 109–113, 1955.
27. A. H. Cottrell and V. Aytekin, *Nature,* **160**, 328, 1947.
28. A. H. Cottrell and V. Aytekin, *J. Inst. Met.* **77**, 389–422, 1950.
29. A. H. Cottrell, *J. Mech. Phys. Solids,* **1**, 53–63, 1952.
30. A. H. Cottrell, in *Creep and Fracture of Metals at High Temperatures,* NPL Symposium, May 1954, Paper No. 8. HMSO, London, 1956.
31. S. Harper and A. H. Cottrell, *Proc. Phys. Soc., Lond.* **B63**, 331–338, 1950.
32. A. H. Cottrell, in *L'Etat Solide*, Neuvième Conseil de Physique, Bruxelles, Septembre 1951, pp. 421–422.
33. H. W. Paxton and A. H. Cottrell, *Acta Met.* **2**, 3–8, 1954.
34. A. H. Cottrell and R. J. Stokes, *Proc. Roy. Soc.* **A233**, 17–34, 1955.
35. R. Maddin and A. H. Cottrell, *Phil. Mag.* **46**, 735–743, 1955.
36. M. A. Adams and A. H. Cottrell, *Phil. Mag.* **46**, 1187–1193, 1955.
37. W. M. Lomer and A. H. Cottrell, *Phil. Mag.* **46**, 711–719, 1955.
38. A. C. Roberts and A. H. Cottrell, *Phil. Mag.* **47**, 1–7, 1956.
39. A. H. Cottrell, in *Dislocations and Mechanical Properties of Crystals*, Int. Conference held at Lake Placid, September 1956, pp. 509–512. Wiley, New York, 1957.
40. A. H. Cottrell, *ibid*, 577–580.
41. A. T. Churchman, I. Mogford and A. H. Cottrell, *Phil. Mag.* **2**, 1271–1275, 1957.
42. R. S. Barnes, G. B. Redding and A. H. Cottrell, *Phil. Mag.* **3**, 97, 1958.
43. A. H. Cottrell, *Metall. Rev.* **1**, 479, 1956.
44. A. H. Cottrell and D. Hull, *Proc. Roy. Soc.* **A242**, 211–213, 1956.

45. A. H. Cottrell, in *UKAEA Conference on Brittle Fracture in Metals*, Culcheth, November 1957, Conf. Prpt./P.P.1.
46. A. H. Cottrell, *Trans. AIME*, **212**, 192–203, 1958.
47. N. J. Petch, in *UKAEA Conference on Brittle Fracture in Metals*, Culcheth, November 1957, Conf. Prpt./P.P.2.
48. A. N. Stroh, *Adv. Phys.* **6**, 418, 1957.
49. A. H. Cottrell, in *Proc. First Tewksbury Symposium*, Melbourne, 1963, 1–27. Butterworths, London, 1965.
50. A. H. Cottrell, in *Conference on Fracture*, Int. Sympm. on Atomic Mechanisms of Fracture, Swampscott, April 1959, 1–1 to 1–33. National Academy of Sciences, USA, 1959.
51. J. Friedel, *ibid*, 18–1 to 18–35.
52. A. H. Cottrell, in *Symposium on Steels for Reactor Pressure Circuits*, London, December 1960, pp. 281–296. Iron and Steel Institute, London, Special Report No. 69, 1961.
53. B. A. Bilby, A. H. Cottrell and K. H. Swinden, *Proc. Roy. Soc.* **A272**, 304–314, 1963.
54. D. S. Dugdale, *J. Mech. Phys. Solids* **8**, 100, 1960.
55. A. A. Wells, in *Proc. Crack Propagation Symposium* **1**, 201. Cranfield, April 1961.
56. F. M. Burdekin, D. E. W. Stone and A. A. Wells, in *Fracture Toughness Testing*, 1965, 400. ASTM STP 381, 1965, 400.
57. B. A. Bilby and J. D. Eshelby, in *Fracture*, an Advanced Treatise (ed. H. Liebowitz), **1**, 99–182. Academic Press, New York, 1968.
58. G. I. Barenblatt, *Prikl. Mat. Mech.* **23**, 434, 706, 893, 1959.
59. G. I. Barenblatt, *Adv. Appl. Mech.* **7**, 55, 1962.
60. P. M. Vitvitskii, V. V. Panasyuk and S. Ya. Yarema, *Engng. Fract. Mech.* **7**, 305, 1975.
61. A. H. Cottrell, in *Properties of Reactor Materials and the Effects of Radiation Damage* (ed. D. J. Littler), Berkeley Conference, June 1961, pp. 5–14. London, Butterworths, 1962.
62. A. H. Cottrell, *Proc. Roy. Soc.* **A276**, 1–18, 1963.
63. A. H. Cottrell, *Proc. Roy. Soc.* **A285**, 10–21, 1965.
64. B. A. Bilby, A. H. Cottrell, E. Smith and K. H. Swinden, *Proc. Roy. Soc.* **A279**, 1–9, 1964.
65. J. R. Griffiths and A. H. Cottrell, *J. Mech. Phys. Solids* **13**, 135–140, 1965.
66. J. F. Knott and A. H. Cottrell, *J. Iron and Steel Inst.* **201**, 249, 1963.
66a. A. H. Cottrell, *The Mechanical Properties of Matter*, Wiley, New York, 1964.
67. A. H. Cottrell, *Proc. Roy. Soc.* **A282**, 2–9, 1964.
68. A. H. Cottrell, in *Metallurgical Achievements*, Session of the Birmingham Met. Soc., 259–272, 1965.
69. A. Kelly, W. R. Tyson and A. H. Cottrell, *Phil. Mag.* **15**, 567–586, 1967.
70. A. Kelly, W. R. Tyson and A. H. Cottrell, *Canad. J. Phys.* **45**, 883–885, 1967.
71. A. H. Cottrell, in *Physics of Strength and Plasticity*, The Orowan 65th Anniversary Volume (ed. A. Argon), pp. 257–261. M.I.T. Press, Boston, 1969.
72. P. T. Heald, G. M. Spink and P. J. Worthington, *Mat Sci. Engng.* **10**, 129, 1972.
73. A. R. Dowling and C. H. A. Townley, *Int. J. Pres. Ves. and Prping* **3**, 77–107, 1975.

74. P. P. Harrison, K. Loosemore and I. Milne, Assessments of the integrity of structures containing defects. CEGB Report R/H/R6, 1976. CEGB, 1976.
75. B. A. Bilby, *Phil. Trans. Roy. Soc.* **A299**, 166, 1981. (Discussion to paper by F. M. Burdekin, *Ibid*, 155–165.)
76. I. Milne, *Mat. Sci. Engng.* **39**, 65–79, 1979.
77. I. Milne, ASTM STP 803, **2**, 657–685.
78. G. C. Chell and I. Milne, ASTM STP 803, **2**, 179–205.
79. I. Milne, R. A. Ainsworth, A. R. Dowling and A. T. Stewart, CEGB Report R/H/R6-Revision 3. CEGB, 1986.
80. I. Milne, R. A. Ainsworth, A. R. Dowling and A. T. Stewart, CEGB Report R/H/.R6-Revision 3, Validation. CEGB, 1987.
81. B. A. Bilby and K. H. Swinden, *Proc. Roy. Soc.* **A285**, 22, 1965.
82. J. R. Rice, *J. Mech. Phys. Solids* **22**, 17, 1974.
83. G. P. Cherepanov, *Prikl. Mat. Mekh.* **40**, 720–728, 1976.
84. C. Atkinson and T. R. Kay, *Acta Met.* **19**, 679, 1971.
85. C. Atkinson and D. L. Clements, *Acta Met.* **21**, 55–60, 1973.
86. J. Weertman, *Proc. Roy. Soc.* **A425**, 291–327, 1989.
87. B. A. Bilby, in *Third International Conference on Fracture*, Munich 1973, Part XI, 1–20. Verein Deutscher Eisenhütten-leute, Dusseldorf, 1973.
88. J. A. H. Hult and F. A. McClintock, in *Proc. 9th Int. Congr. Appl. Mech.* **8**, 51–58, University of Brussels, Brussels, 1957.

2 Theory of the Yield Point and of Strain-Ageing in Steel

N. J. Petch

Emeritus Professor of Metallurgy, University of Strathclyde

All of us know that science advances principally by small steps in which previous ideas are taken a little bit further. However, now and then, there is a real stride forward; a stride that opens up whole new pastures for thought, understanding and experiment. Cottrell's idea of carbon or nitrogen locking of dislocations was such a step in the metallurgy of steel.

Initially, dislocation theory seemed quite interesting for simple-minded metals, such as copper or aluminium, but there was not much impact on the real stuff — steel. Suddenly, kicking in the front door of steel metallurgy, came Cottrell's idea. Here was theory brazenly talking about the yield point, something in which steel had a real proprietorial interest. Attention had to be given, and dislocation theory, not only burst in, but, came to stay.

It is always of interest to know how ideas have arisen. In my own case, such ideas as I ever had, seem to have random-walked into my mind whilst I was shaving. The advent of the electric shaver has rather spoilt that. I am pleased to say that Sir Alan's approach was much more scientific. I have it straight from himself. Nabarro's treatment of solid solution hardening, in which a dislocation is elastically repelled by solute atoms, led Alan in 1946 to think of a 'snow plough' model in which a moving dislocation would push the solute atoms out of the way and clear a channel for other dislocations, so giving a yield drop. He then realised that in other parts of the dislocation stress field, solute atoms would be attracted, not repelled, and much more significant effects could arise. These thoughts were rather laid aside because of experimental work, but the fuel crisis of 1947 brought that to a halt, and he then had time to explore the calculation of the interaction energy between the dislocation and solute atom. Concurrently, he had realised from the literature that the steel yield point was associated with the presence of small amounts of C and N. So, hey presto!

Cottrell[1] considered the attraction between a solute atom and an edge dislocation in terms of the relief of the hydrostatic stress at the dislocation.

This gave a binding energy

$$U = p\,\Delta\upsilon,$$

where p is the hydrostatic stress and $\Delta\upsilon$ the volume change produced by the solute atom.

If the solute atom is at r, θ with respect to a positive edge dislocation,

$$U = \frac{A \sin\theta}{r} \tag{1}$$

where A depends on the elastic constant, the volume change and the Burgers vector. In the case of C or N in steel, U increases so much as r falls that condensation of the atmosphere was proposed, giving a line of solute atoms at $r \sim 2 \times 10^{-8}$ cm and $\theta = 3\pi/2$.

With a pure screw dislocation, there is no hydrostatic stress, so this type of interaction cannot occur. However, C and N atoms in steel produce a tetragonal distortion and this interacts with the shear stress of the screw dislocations about as strongly as the volume change interaction with the edge dislocations.[2]

The calculated value of the stress to free a dislocation from a C–N atmosphere is a number of times the yield stress at absolute zero. This suggests some form of stress concentration may be needed to achieve dislocation movement. Thermally-activated throwing forward of a dislocation loop was analysed by Cottrell and Bilby;[3] this lowers the stress required for freeing a dislocation at a higher temperature, but the yield stress is also lowered. Thermally activated unlocking has also been analysed by Fisher, Louat and Lothe.[3]

I first heard of Cottrell's ideas from Orowan when he came back from some meeting and I well remember being rather puzzled because I knew the yield point disappeared at coarse grain sizes, which did not seem to fit in with these ideas. This was a valid point, but all it shows is that you cannot expect to deal with everything at one go.

With help from metallographic studies[4] and from analysis of the stress at the head of a pile-up of dislocations[5] the following picture of the yield point in steel has emerged.

Initially, all the dislocations are locked. On straining, some dislocations are eventually released at regions of stress concentration, such as fine particles, grain boundary ledges or triple points. The dislocations spread across randomly distributed grains, but are held up at the grain boundaries (pre-yield microstrain). When a cluster of such grains forms of sufficient size, the macroscopic stress concentration it produces results in grain boundary break-through, and the plastic zone spreads across the specimen. The upper yield point has then been achieved. Subsequently, the yielded zone, now termed a Luders band, spreads along the specimen at the lower yield point. The dislocation multiplication between the upper and lower yield points means that slower dislocation movement can achieve the applied strain rate, so the stress falls.

Unless extreme care is taken to ensure axial loading, measurements of the upper yield point are uncertain, so attention has concentrated on the more easily measured lower yield point. The argument that has been developed is that propagation of the Luders band occurs by virtue of the stress concentration generated by blockage of slip bands at the grain boundaries between the yielded and unyielded zones. The stress generated depends on the length of the slip band and on the applied stress less any friction stress on the dislocations. This leads to a relationship for the lower yield stress,

$$\sigma_y = \sigma_o + k_y d^{-\frac{1}{2}}, \tag{2}$$

where σ_o is the in-grain friction on the dislocations and $k_y d^{-\frac{1}{2}}$ measures the resistance of the grain boundary to slip transmission, with k_y a constant and d the grain size.[6]

This simple relationship, directly dependent on Cottrell's recognition of dislocation locking, brought a great deal of understanding to the mechanical properties of steel. Its great merit was that it identified and separated the various factors that determine the yield stress and permitted them to be studied individually.

A point of major interest was the realisation for the first time that in α-iron, unlike f.c.c. metals, there is a major, temperature and strain-rate dependent contribution to the dislocation friction σ_o from the intrinsic structure resistance (the Peierls–Nabarro stress). Substitutional solutes can affect this, producing alloy-hardening at room temperature, but the possibility of alloy-softening at lower temperatures.[7]

Precipitations and substructures make a contribution to σ_o, and this can be separated, since it is athermal.[8] Irradiation damage also increases σ_o.[8]

Equation 2 brought realisation of the great potential benefit of grain refinement and put this on a quantitative basis.

Initially, it was thought that k_y was determined by the unlocking of a dislocation from its atmosphere in the grain ahead of the blocked slip band. Now, as discussed later, it is considered that, for an annealed steel, dislocations are derived from the grain boundaries that blocks the slip band. The stress required is influenced by the segregation of interstitials to the boundary. There is normally more nitrogen than carbon in solution, so additions such as Si, Mn or Al, that interact with nitrogen, affect boundary segregation and so affect k_y.[9] Thus, k_y varies over the range ~17–23 $\mathrm{Nmm}^{-3}/2$ depending upon deoxidation procedure.

The value of k_y is temperature-independent down to about −150 °C, but then increases.[10] Possibly this increase reflects a change to planar slip and a consequent change in the orientation factors involved in the polycrystalline deformation rather than a change in the concentrated shear stress required for grain boundary transmission.[6, 36]

The σ_y relationship has been used, for example, by Irvine, Gladman and Pickering *et al.* in many papers examining the relationship between mechanical properties and microstructure of steels.[11] With ferritic steels, the analysis is clear. With bainitic steels, there are so many sources of dislocation-unlocking that there is no sharp yield point, but the proof stress is accountable in terms of a $d^{-\frac{1}{2}}$ dependence on the bainitic packet size (involving large angle boundaries), with σ_o influenced by in-grain lath and other structures.[12] With martensite, amongst other factors, there appears to be a martensite packet $d^{-\frac{1}{2}}$ contribution. Tempering bainite or martensite adds further complication, but basically there are barriers and their penetration by pile-ups is involved.

Thirty years ago, higher carbon contents (0·4 C) were used for the higher yield strength steels required in many important engineering applications. However, these gave high ductile-brittle transition temperatures and cracking problems in welding. A major development has been their replacement by low carbon (0·03/0·05) microalloyed steels (HSLA steels). Equation 2 has guided the development of these.[13] Fine grain sizes ($d^{-\frac{1}{2}} = 10\,\mathrm{mm}^{-\frac{1}{2}}$) are achieved by controlled rolling in the austenitic region. Further refinement results with alloy additions (<0·1%) involving Nb, Ti, Va and Al that produce carbide, nitride or carbonitride particles. There are also increases in σ_o due to these precipitates, dislocations and substructures. Such microalloyed steels can give yield points up to 650 Nmm^{-2} compared to the 200–250 Nmm^{-2} of plain carbon mild steel.

A closer look at the stress drop between the upper and lower yield points is of interest. This drop arises from an increase in the density of mobile dislocations and a consequent decrease in the effective strain rate on them. Since the stress to move a dislocation in α-iron does not change rapidly with strain rate, and approximate treatment of the stress-drop can be attempted.[14]

The stress–strain curve flattens as the upper yield point, σ_{uy}, is approached, indicating that the strain is then predominantly pre-yield plastic deformation with little elastic contribution. Metallographic evidence shows that the number of pre-yielding grains, N per mm^3, is fairly independent of grain size.[4, 15] Thus, the deforming volume per unit specimen volume is Nd^3 and the strain rate in these grains is inversely proportional to this. If the dislocation structure in the deforming grains is reasonably independent of grain size, the friction stress $\sigma_{o(u)}$ in them at the upper yield point will then vary with grain size as

$$\sigma_{o(u)} = \text{constant} + \Delta\sigma_o \log 1/Nd^3, \tag{3}$$

where $\Delta\sigma_o$ is the increase in friction stress due to a tenfold increase in strain rate.

At coarse grain sizes, $\sigma_{o(u)}$ falls to the constant value corresponding to $Nd^3 = 1$, (i.e.) all the grains deform in preyield. Then both the upper and lower yield points will disappear and the deformation is homogeneous.

Thus, provided the stress for homogeneous deformation is near to the lower yield stress value expected for that $d^{-\frac{1}{2}}$,

$$(\sigma_{uy} - \sigma_{ly}) = \Delta\sigma_o \log 1/Nd^3.$$

Values of $(\sigma_{ly} + \Delta\sigma_o \log 1/Nd^3)$ are shown in Fig. 1 for three values of N, along with mild steel experimental points for σ_{uy}. There is reasonable agreement at $N = 10/mm^3$.

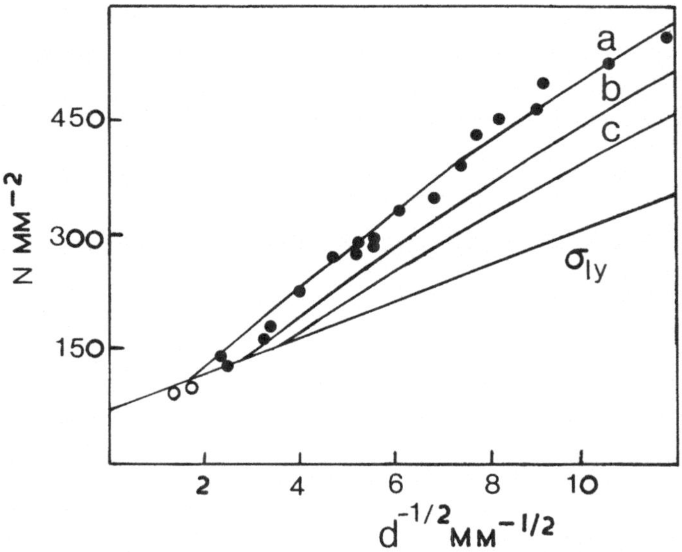

Fig. 1. Experimental σ_{ly} for a mild steel at 18 °C (homogeneous yielding ○), with $(\sigma_{ly} + \Delta\sigma_o \log_{10} 1/Nd^3)$ for N equals (a) 10 (b) 10^2 (c) 10^3. The experimental points ● are for σ_{uy}.[14]

It will be noted that this explains the disappearance of the upper and lower yield points at coarse grain sizes, which was the point that puzzled me when I first heard of Cottrell's theory.

With 3% Si-Fe, the yield drop may involve additional complications.[15]

Single crystals may show a yield drop. Again, there is localised dislocation unlocking and yielding spreads through the crystal, the stress falling as the mobile dislocation density increases. However, the definite boundaries to spread that a polycrystal presents are absent, and with that the well-defined lower yield point strain.

Turning to the flow stress after σ_{ly}, this also shows a $d^{-\frac{1}{2}}$ dependence, but with a slope k_ϵ about half k_y. The value of k_ϵ may again involve the grain boundary resistance to the transmission of slip, but this probably depends on the operation of unlocked sources now available ahead of the blocked slip band. In addition, any grain size dependence of the work-hardening rate will contribute to k_ϵ.

In the Luders band, the grains at the interface with the unyielded zone contain few, principally parallel, slip bands, so there is no work-hardening contribution to σ_o. Behind this front, unlocked dislocations become available; ordinary multiple slip occurs and there is strain hardening of the flow stress for these grains up to σ_{ly}.[4]

At the end of the Luders strain, the flow stress equals σ_{ly}. Thus,

$$\sigma_o + k_y d^{-\frac{1}{2}} = \sigma_o^1 + \text{strain hardening} + k_\epsilon d^{-\frac{1}{2}},$$

where σ_o^1 is the initial friction stress in homogeneous deformation. In steel, the strain-hardening rate during homogeneous deformation is not linear, but approximates to this over small strains. With this, and taking $\sigma_o^1 \sim \sigma_o$, the Luders strain ϵ_L is

$$\epsilon_L = \text{constant}\,(k_y - k_\epsilon)d^{-\frac{1}{2}}.$$

Over the common range of grain sizes, this $d^{-\frac{1}{2}}$ dependence is reasonably true (Fig. 2).

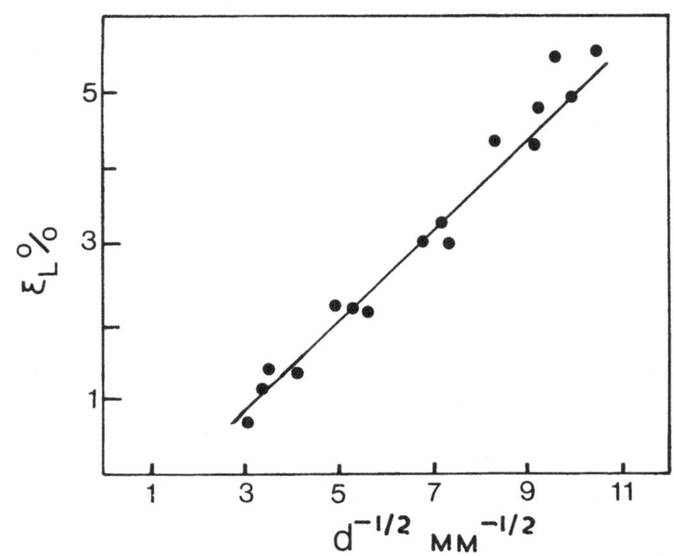

Fig. 2. The d$^{-\frac{1}{2}}$ *dependence of the Luders strain for a mild steel at 18°C.*

Static Strain-Ageing

The free dislocations generated by straining can be relocked by diffusion of the interstitials to them. After a preliminary analysis of the rate of this process by Nabarro,[2] the classical treatment was given by Cottrell and Bilby.[3] It indicated a time 2/3 dependence. This considered only the drift of the interstitials to the dislocation and referred only to the earlier stage of the process. Experimentally, there is agreement with this. After 30–40% of the ageing, there is complication from back diffusion and from precipitation on the dislocations.[16]

A pleasing association with Cottrell is the extensive work done by Wilson and co-workers and by others associated with Birmingham — Cottrell's home University.

The effect of ageing on σ_o and k_y was examined. As the ageing proceeds, σ_o for the returned yield point increases; k_y also initially increases rather quickly, but then reaches a plateau value, less than k_y for the original annealed material. On more prolonged ageing, there is a slow rise in k_y up towards the annealed value (Fig. 3).

The increase in σ_o reflects in-grain hardening due to atmosphere and

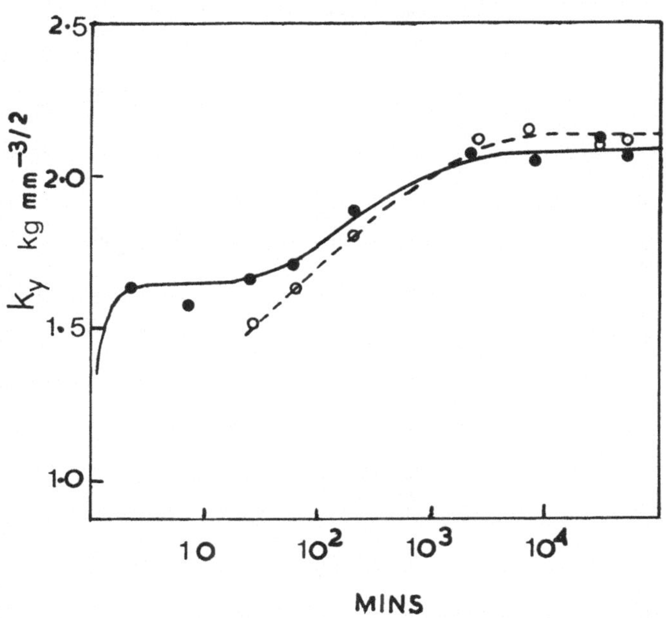

Fig. 3. The change in k_y *with ageing time for a 0·04 °C steel, quenched from 600 °C before 5% prestrain. Longitudinal tests ●, transverse ○.*[21]

precipitate formation. The interpretation put on the k_y changes is that, in the earlier stages of ageing, yield is propagated by the operation of lightly-locked dislocation in the grain ahead of the blocked slip band. The k_y value then increases as the locking becomes progressively stronger. The occurrence of the plateau is interpreted as a transition. Instead of continuing to operate a source in the next grain, it becomes easier to derive a dislocation from the grain boundary that is blocking slip. The strain prior to ageing modifies the grain boundaries, and the slow increase from the plateau value of k_y is considered to represent the slow re-establishment of the equilibrium interstitial segregation to the grain boundaries.[17]

A similar conclusion was reached from quench-ageing measurements by Cottrell and Fisher.[18] Figure 4 shows k_y values after various times of ageing at 140 °C. This is taken to indicate that, after short ageing, yielding is propagated by the temperature-dependent unlocking of atmosphere-locked, in-grain dislocations. As the temperature is lowered, this unlocking becomes more difficult and there is a transition to a process giving a temperature-independent k_y. On longer ageing, k_y at room temperature is raised, but again, at lower temperature, there is a transition to the same temperature-independent value.

Interesting transmission electron microscopy evidence on this point came from a study of strain-ageing in vanadium.[19] In the early stages of

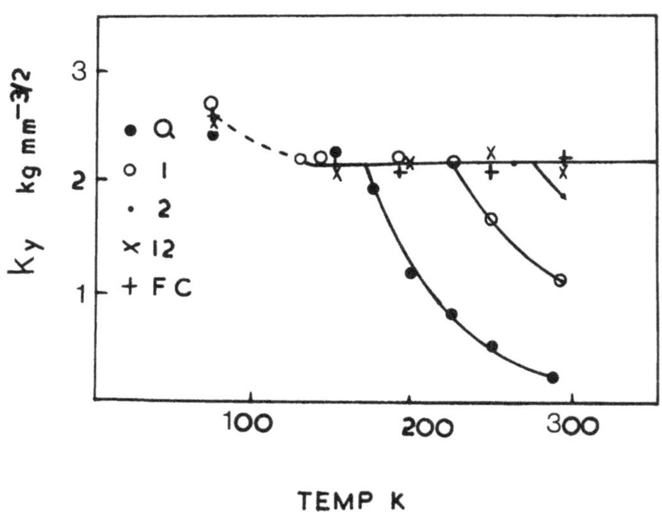

Fig. 4. The dependence of k_y *on temperature for iron after quenching and various ageing treatments or after furnace cooling. (• quenched; aged at 140 °C for (○) 1, (•) 2 or (X) 12 h, or (FC) furnace cooled).*[18]

ageing, the dislocations in the slip bands formed on restraining had the same Burgers vector as the prestrained dislocations. This was interpreted as indicating that, on restraining, in-grain dislocations were unlocked and continued their movement. However, in the later stages of ageing, restraining introduced a new set of dislocations with a different Burgers vector. This was taken to indicate the operation of a new source.

Directional effects arise during restraining a prestrained steel.

First, there is the Bauschinger affect arising from internal stresses generated by the prestrain. Soft grains are left in compression; hard grains in tension. If restraining is in the original prestrain direction, soft and hard grains will deform at about the same external stress. However, if the restraining is in the reverse direction, the internal stress will favour early deformation of the soft grains. Thus, reverse deformation will start at a low external stress, but will rapidly harden as the internal stress is relaxed.

Secondly, there is a directionality in the return of the sharp yield point. This was observed by Mrs. Tipper[20] and examined by Wilson, Ogram and Bate.[21, 35] If the direction of restraining differs from the prestrain direction, the initial return of the yield point is considerably slower than if both directions are the same. At this stage, the locking of in-grain dislocations is involved. It is suggested that a polarity in the distribution of the relevant dislocations explains the directional effect. When positive dislocations in grain A pile up against a grain boundary, the near-boundary dislocations in the next grain B are predominantly −ve, the +ve dislocations will be fewer and more distant. On restraining in the same direction these more-distant +ve dislocations are used in the propagation of slip. However, if the direction of restraining is reversed, +ve dislocations will now arrive in grain B at this boundary, and propagation will be by positive dislocations in grain A. The prestrain will have produced a dense pack of these at the grain boundary. It is then argued that atmosphere-locking of this dense pack takes a long ageing time, whereas the locking of the few, more isolated, relevant dislocations, in the case of restraining in the original direction, is quicker.

Even when the sharp yield point has returned in reverse restraining, it is at a lower stress than in forward restraining. This is because of the persistence of the Bauschinger internal stresses.

On prolonged ageing, k_y eventually depends on dislocations derived from the grain boundary, the direction of restraining does not affect the ageing time required for this (Fig. 3).

Industrially, Luders bands (stretcher-strains) are of importance in the use of sheet and strip for cold-worked produces; the bands spoil the surface appearance. This can be overcome by extending through the yield strain before use. However, that may well require a strain of ~5%, and the yield point returns in the stretch direction in a day or so. In industry, the annealed sheet or strip is 'temper-rolled' to about 1½% reduction.[22] It then takes a number of weeks for the return of a sharp yield point. Temper-rolling produces a structure of interwoven yielded and unyielded zones. On straining after a few days ageing, deformation spreads into the unyielded zones on numerous fine Luders bands along the interfaces between the two zones, and an acceptable surface appearance results. After more prolonged ageing, the number of centres of deformation propagation decreases and eventually is reduced to just a few. There is then a return of a sharp yield point. The delay in this return compared to that in simple pre-stretching appears to arise from the directional effects in restraining already discussed.

Dynamic Strain-Ageing

Raising the test temperature produces a marked change in the stress–strain curves for steel.

At 100 °C, jerky flow develops after some strain; at 150 and 200 °C, there are marked jerks even in the yield point extension. At these temperatures, there is a strain-hardening rate increase and a reduction in the

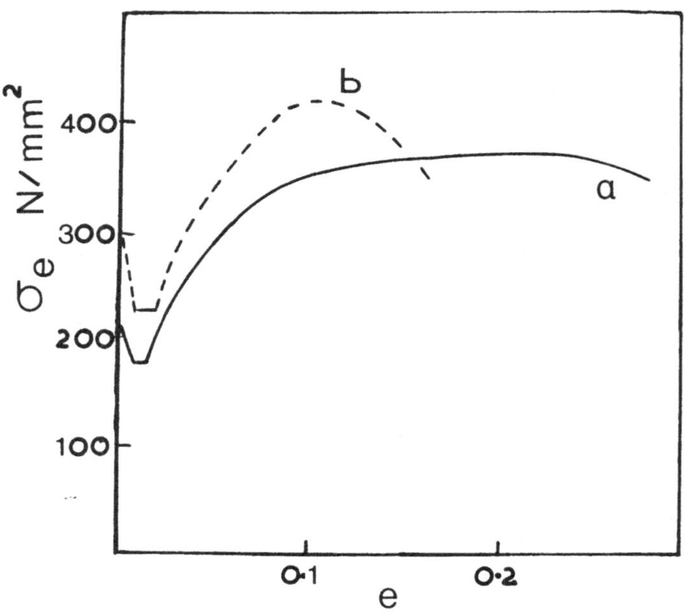

Fig. 5. Dynamic strain-ageing of a mild steel. Nominal stress vs elongation at (a) room temp. (b) 150 °C.[23]

ductility ('blue brittleness'). This is shown in Fig. 5 with the jerks smoothed out. By 300 °C, the jerks have gone; the strain-hardening rate remains high, but drops with further rise in temperature.

These effects arise because strain-ageing occurs during plastic deformation at these temperatures.

Again, Cottrell was involved in the initial basic consideration of this matter. He and Jaswon[24] analysed the interaction between a solute atom and a moving dislocation (see also Hirth and Lothe[25]).

At room temperature, the mobility of C or N atoms is too low for them to move with a dislocation. However, at not much higher temperatures, they can trail along, dragged by interaction with the dislocations.

In a simplified version,[26] if a force F acts on an atom migrating by thermal agitation, the atom develops a drift velocity v in the direction of F such that $v = (D/kT)F$, where D is the diffusion coefficient.

For an edge dislocation interacting with a solute atom at a distance r, this gives $v = (D/kT)(A/r^2)$, where A is the interaction parameter in equation 1.

A/kT has the dimensions of length and can be regarded as a measure of the radius ℓ of the atmosphere; it is smaller, the higher the temperature. Taking an atom at $r = \ell/2$ as representative, the atomic velocity is $4D/\ell$. This then is the critical velocity for the dislocation; if it moves faster, the atmosphere is left behind; if slower, the atmosphere can keep up. With a dislocation density ρ, the critical shear strain rate is

$$\frac{d\gamma}{dt} = v\rho b = 4D\rho b/\ell \tag{4}$$

Thus, increasing the temperature gives a smaller, faster moving atmosphere that can move with the dislocation at a higher strain rate.

The dragging force on the dislocation due to an average atom in the atmosphere at $r = 1/2$ is $4A/\ell^2$. For the whole atmosphere, this gives a dragging stress of $\sim 28Ac_oN/b$, where c_o is the average atomic concentration of the solute and N is the total number of all atoms per unit volume.

For the normal diffusion rate of substitutional solutes, equation 4 indicates it will only be in creep that the solute can move with the dislocations. In this way, substitutional atoms (Mn, Cr, Mo etc.) in steel increase the creep resistance in the 400–500 °C range.[32] However, drag of substitution atoms does occur, not only in creep, but also in ordinary tensile tests (e.g. α-brass).[27] As Cottrell[29] has pointed out, this happens only after strain has generated vacancies and so enhanced D.

The jerky flow associated with dynamic strain-ageing arises because the dislocations do not move with a steady velocity; they are liable to be held up, e.g. at grain boundaries, and this permits the solute atmosphere to catch up. Thus, periodically, there are lockings, and a raised stress is required to restart deformation. In the Luders strain, the repeated arrest of the Luders band front shows up in the surface markings and the stress rises practically to the original upper yield point. Downward jerks also occur associated with variations in the atmosphere drag stress due to variations in dislocation velocity.[37]

In steel, nitrogen atoms may cluster in solution with substitutional ones, e.g. Mn; this slows down their rate of diffusion and so displaces these dynamic strain-ageing effects to higher temperatures.[29]

The increase in the flow stress and strain-hardening rate during dynamic ageing arises partly from the increase in dislocation density due to locking (Fig. 6). Increased generation of fresh dislocations is required to maintain deformation; also there is reduced dislocation annihilation.[30] Precipitates and atmosphere-drag also contribute to the hardening.

In this dynamic strain-ageing temperature range, there is the unusual effect that increasing the strain rate lowers the flow stress. This is, of course, simply because there is less ageing.

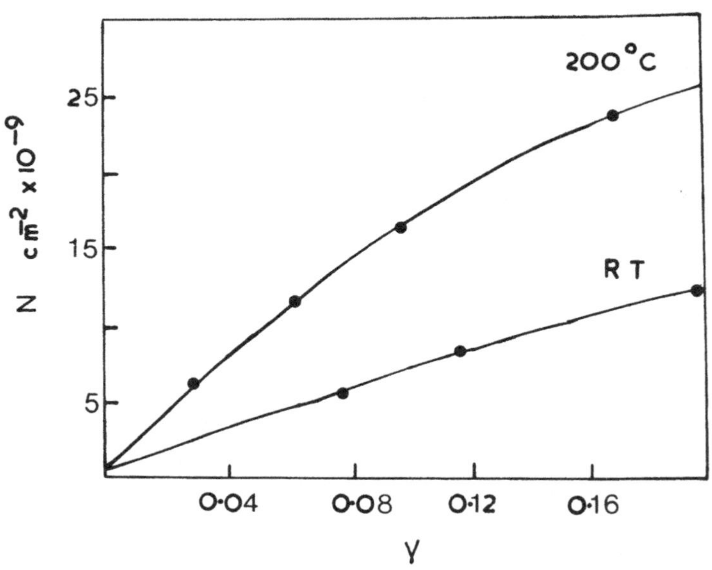

Fig. 6. Dislocation density N *vs plastic shear strain* γ *for a 0·035 °C steel deformed at (a) room temp. (b) 200 °C.*[30]

Blue brittleness is sometimes commented on as if it were simply the result of the increased strain-hardening rate during dynamic strain-ageing (Fig. 5), but further consideration is needed. The onset of necking occurs when the rate of strain-hardening equals the flow stress. An increase in hardening rate increases both of these. The important point in blue brittleness is that the initial increased hardening rate does not persist at higher strains when dislocation locking has reduced the availability of interstitial atoms for strain-ageing. Thus, the flow stress is raised by the initial hardening, but the hardening rate falls back and the strain for the onset of necking is reduced from ~30% to ~10%, with a consequent reduction in the elongation to fracture. Baird and Jamieson[31] observed that the loss in ductility does not occur with high nitrogen contents (>0·03%).

Fatigue

Dislocation-locking comes up again in the fatigue of steel. In contrast to most other engineering metals, steel shows a fatigue limit — a stress below which fatigue cracking does not develop. It was thought that dynamic strain-ageing was responsible for this, but it appears that the spread of plasticity across grain boundaries is the important factor.[33] In fatigue just below the limit, for all but very coarse grain sizes, only a small proportion of the surface grains deform plastically; this deformation cannot spread into neighbouring grains and fatigue cracks do not develop. Above the limit, spread of plasticity and fatigue cracking occurs. The fatigue limit shows a $d^{-\frac{1}{2}}$ dependence, but the slope is less than the normal tensile k_y. This suggests a change in the grain boundaries by the cyclic stressing, or a change to the operation of in-grain dislocations in the grain boundary transmission. On the contribution of dynamic strain-ageing, it is noted that a fatigue limit can occur at temperatures too low (−60 °C) for dynamic ageing, but at higher temperatures, when ageing is possible, the limit is raised.[39]

In the extension of pre-existing cracks, slip band penetration of grain boundaries is again important. If the plastic deformation ahead of the crack is blocked by a boundary, the held-up dislocations exert a back stress on the crack, inhibiting, and in the extreme case, stopping, growth. This appears to be the reason for the existence of the threshold stress intensity.[34]

Conclusion

I hope this review has brought out just how many theoretical and industrial developments in this aspect of steel metallurgy have their roots back in Cottrell's theory of dislocation locking. I wonder how far we would have got without it. I wonder also if there is a lesson here to be pondered by those currently so keen on developing central bureaucratic control on planning and funding of research in the Universities of this country.

Would they, 40 years ago, have recognised the importance for steel metallurgy of research on the interaction of a solute atom with the stress field of a dislocation?

References

1. A. H. Cottrell, *Report on the Strength of Solids*, London, Phys. Soc., 30, 1948.
2. F. R. N. Nabarro, *Ibid*, 38, 1948.
3. A. H. Cottrell and B. A. Bilby, *Proc. Phys. Soc.*, **A62**, 49, 1949. J. C. Fisher, *Trans. ASM.*, **47**, 451, 1955. N. Louat, *Proc. Phys. Soc.*, **B69**, 454, 1956; **B71**, 144, 1958. J. Lothe, *Acta metall.*, **10**, 663, 1962.
4. D. F. Stein and J. R. Low, *J. appl. Phys.*, **31**, 362, 1960. J. Holden, *Acta metall.*, **8**, 424, 1960. J. C. Suits and B. Chalmers, *Ibid*, **9**, 854, 1961. P. J. Worthington and E. Smith, *Ibid*, **12**, 1277, 1964. W. E. Carrington and D. McLean, *Ibid*, **13**, 493, 1965. W. B. Morrison and R. C. Glenn, *J. Iron Steel Inst.*, **206**, 618, 1968. W. D. Brentnall and W. Rostoker, *Acta metall.*, **13**, 187, 1965.
5. C. Zener, *Phys. Rev.*, **69**, 128, 1946. J. D. Eshelby, F. C. Frank and F. R. N. Nabarro, *Phil. Mag.*, **42**, 351, 1951.
6. E. O. Hall, *Proc. Phys. Soc.*, **B64**, 742, 747, 1951. N. J. Petch, *J. Iron Steel Inst.*, **174**, 25, 1953. R. W. Armstrong, I. Codd, R. M. Douthwaite and N. J. Petch, *Phil. Mag.*, **7**, 45, 1962.
7. W. C. Leslie, R. J. Sober, S. G. Babcock and S. J. Green, *Trans ASM.*, **62**, 690, 1969.
8. A. Cracknell and N. J. Petch, *Acta metall.*, **3**, 186, 1955. A. H. Cottrell, Iron Steel Inst., Special Report No. 69, p. 281, 1961. D. Hull and I. L. Mogford, *Phil. Mag.*, **3**, 1213, 1958.
9. I. Codd and N. J. Petch, *Phil. Mag.*, **5**, 30, 1960. G. Birkbeck and N. J. Petch, *J. Iron Steel Inst.*, **207**, 607, 1969. N. J. Petch, *Acta metall.*, **35**, 2027, 1987. B. Mintz, *J. Iron Steel Inst.*, **211**, 433, 1973.
10. N. J. Petch, *Phil. Mag.*, **3**, 1089, 1958.
11. K. J. Irvine, F. B. Pickering and T. Gladman, *J. Iron Steel Inst.*, **205**, 161, 1967. T. Gladman and F. B. Pickering, *'Yield, Flow and Fracture of Polycrystals'* (ed. T. N. Baker). London, Applied Science Publishers, p. 141, 1983.
12. R. W. Honeycombe and F. B. Pickering, *Met. Trans.*, **3**, 1099, 1972.
13. W. E. Duckworth, R. Phillips and J. A. Chapman, *J. Iron Steel Inst.*, **203**, 1108, 1965. W. B. Morrison and J. H. Woodhead, *Ibid*, **201**, 43, 317, 1963. A. J. Lapointe and T. N. Baker, *Met. Sci. J.*, **16**, 207, 1982. T. N. Baker, *Science Progress.* Oxford, **65**, 493, 1978.
14. N. J. Petch, *Acta metall.*, **12**, 59, 1964.
15. P. J. Worthington, *Ibid*, **15**, 1795, 1967.
16. R. Bullough and R. C. Newman, *Proc. Roy. Soc.*, **A249**, 427, 1959; **A266**, 178, 209, 1962. B. A. Bilby, *J. Iron Steel Inst.*, **184**, 65, 1956. F. S. Ham, *J. appl. Phys.*, **30**, 915, 1959.
17. D. V. Wilson and B. Russell, *Acta metall.*, **8**, 36, 1960. D. V. Wilson, *Met. Sci. J.*, **1**, 40, 1967; *Acta metall.*, **16**, 743, 1968. W. C. Leslie and A. S. Keh, *J. Iron Steel Inst.*, **200**, 722, 1962.

18. See A. H. Cottrell, *'Relation between Structure and Mechanical Properties'*, p. 445. London, N.P.L., 1963.
19. J. W. Edington, T. C. Lindley and R. E. Smallman, *Acta metall.*, **12**, 1025, 1964.
20. C. F. Tipper, *J. Iron Steel Inst.*, **172**, 143, 1952.
21. D. V. Wilson and G. R. Ogram, *Ibid*, **206**, 911, 1968.
22. B. B. Hundy, *Ibid*, **181**, 313, 1955. R. D. Butler and D. V. Wilson, *Ibid*, **201**, 16, 1963.
23. B. J. Brindley, *Acta metall.*, **18**, 249, 1955.
24. A. H. Cottrell and M. A. Jaswon, *Proc. Roy. Soc.*, **A199**, 104, 1949.
25. J. P. Hirth and J. Lothe, *'Theory of Dislocations'*, McGraw-Hill, 1968.
26. A. H. Cottrell, *'Dislocations and Plastic Flow in Crystals'*, p. 136. Clarendon Press, 1953.
27. G. W. Ardley and A. H. Cottrell, *Proc. Roy. Soc.*, **A219**, 328, 1953.
28. A. H. Cottrell, *Phil. Mag.*, **44**, 829, 1953.
29. J. D. Baird and C. R. Mackenzie, *J. Iron Steel Inst.*, **202**, 427, 1964.
30. D. J. Dingley and D. McLean, *Acta metall.*, **15**, 885, 1967. A. S. Keh, Y. Nakada and W. C. Leslie, *'Dislocation Dynamics'*, p. 381. McGraw-Hill, 1968.
31. J. D. Baird and A. Jamieson, *J. Iron and Steel Inst.*, **204**, 793, 1966.
32. J. Glen, *Ibid*, **190**, 114, 1958. J. D. Baird and A. Jamieson, *Ibid*, **210**, 841, 847, 1972.
33. D. V. Wilson and G. Oates, *Acta metall.*, **12**, 21, 1964. B. Mintz and D. V. Wilson, *Ibid*, **13**, 947, 1965.
34. A. Navarro and E. R. de los Rios, *Phil. Mag.*, **A57**, 15, 1988.
35. P. S. Bate and D. V. Wilson, *Acta metall.*, **34**, 1097, 1107, 1986. D. V. Wilson and P. S. Bate, *Materials Forum*, **10**, 33, 1987.
36. E. Smith and P. J. Worthington, *Phil. Mag.*, **9**, 211, 1964.
37. F. R. N. Nabarro, *Theory of Crystal Dislocations*, p. 431. Oxford, Oxford University Press, 1967. A. Van den Beukel and U. F. Kocks, *Acta metall.*, **30**, 1027, 1982

3 Ductile and Brittle Failure in Steel/ Pressure Vessel Assessments

F. M. Burdekin

Professor, Department of Civil and Structural Engineering,
University of Manchester Institute of Science and Technology

Abstract

The current position with respect to application of fracture mechanics methods to assessment of failure in steel structures is discussed with particular reference to the safety of nuclear pressure vessels. The implications of such assessments on non-destructive testing requirements are considered in the light of current developments of techniques and validation.

It is concluded that with modern materials, fabrication practices and diverse and redundant inspection techniques, acceptably safe construction can be achieved. The basic understanding of fracture mechanisms is a key requirement to confidence in assessment procedures, and to the development of further improvements in materials, and quality control test methods.

Introduction

Extensive research has been carried out over the past thirty years or so into improving understanding of detailed mechanisms of ductile and brittle failure in steels, and into methods of assessment of practical design and construction to avoid such failures in service. Of course it is not possible to guarantee absolute security against failure, since any material will fail if forces are applied which exceed its resistance. The basic requirement is to achieve a level of security which is acceptable to human society. One way of quantifying this is to determine the probability of a catastrophic failure, and to determine its consequences in terms of the potential number of fatalities which might result, with the requirement that the product of probability and consequences should not exceed a particular constant number. The choice of this constant will certainly be different for cases where the general public has some choice over activities (e.g. motor cars, smoking) and cases where it has no choice in industrial activities (e.g. industrial plant).

One of the most critical applications where it is essential that risks of failure by fracture are kept to very low and acceptable levels is in the nuclear power industry. This problem is reviewed by Sir Alan Cottrell in his book *How Safe is Nuclear Energy*.[1] Cottrell expresses the view 'that he does not think the coming multitudes can survive the next century without it. Nuclear energy can be made sufficiently safe, certainly much safer than any realistic large scale alternative. Whether it will be safe in practice is an entirely different matter which depends on making the right detailed choices.'

With respect to the Pressurised Water Reactor (PWR) Cottrell states that security of the pressure vessel is of over-riding importance.[1] He expresses the view that essential steps for the safety of such pressure vessels are as follows:

(i) apply rigorously all recommendations of the Marshall Report for manufacture, operation and inspection of the vessels,
(ii) improve ultrasonic techniques and procedures to the level at which in routine practice they detect 6 mm cracks with fair probability, and 25 mm cracks with very high probability,
(iii) develop methods for repair remotely in thick radioactive steel without impairing mechanical properties,
(iv) repair by these methods all cracks deeper than 6 mm.

Cottrell concludes that, 'the long thrust of life over the millenia is inevitably towards any new energy source which will succour it. The next move into nuclear energy is inevitable. We should not use our fears to try to turn the clock back, but instead use our intelligence in order to enjoy the benefits of nuclear power safely.'

Clearly to be able to quantify the risks of fracture in steel it is necessary to understand the detailed mechanisms and to have assessment procedures which model structural behaviour adequately.

Fracture Mechanisms

Brittle fracture by cleavage

A major contribution to the understanding of mechanisms of cleavage fracture was made by Cottrell in a series of papers between 1957 and 1964[2–9] with independent proposals involving some similar concepts from Petch[10, 11] and Wells.[12] The basic concept put forward by Cottrell was that cleavage fracture occurred by the coalescence of glide dislocations to form microcracks, the growth of which required a stress higher than that for coalescence. This led to the concept of a maximum displacement between faces of a crack in terms of a pile up of dislocations (subsequently to become recognised as crack opening displacement) and the concept of a line integral around the crack tip from one face to the other leading to work done on dislocations equalling the surface energy of crack faces opened by movement (similar to J-contour integral). Among attempts to produce experimental evidence of the coalescence of dislocations to produce microcracks was that of Davies and Wells.[13] The source of initiating microcracks for cleavage fracture was extended to include cracking of inclusions in the work of Smith[14, 15] with an analysis of conditions to propagate microcracks into the surrounding matrix.

The work on the stress level to cause propagation of microcracks was continued by Knott and Cottrell,[9] Knott,[16] and Ritchie, Knott and Rice[17] leading to the conclusion that cleavage cracks propagate when the maximum local normal stress exceeds a critical fracture stress over a critical distance ahead of the crack tip, the critical fracture stress being relatively independent of temperature.

These concepts have been re-examined recently in work at UKAEA Harwell by Gibson.[18] In detailed fractographic work on cleavage fractures in A508 (Class 3) steel Gibson found that cleavage fracture was initiated by cracking of brittle second phase particles, namely iron carbides, titanium carbides, or manganese sulphide inclusions. Gibson considered the previous analyses of Smith[14, 15] and the crack tip stress analyses of McMeeking[19] and of Griffiths and Owen[20] and concluded that under elastic plastic conditions the position of the peak tensile stress moves further away from the crack tip with increased loading (Fig. 1). Thus at a

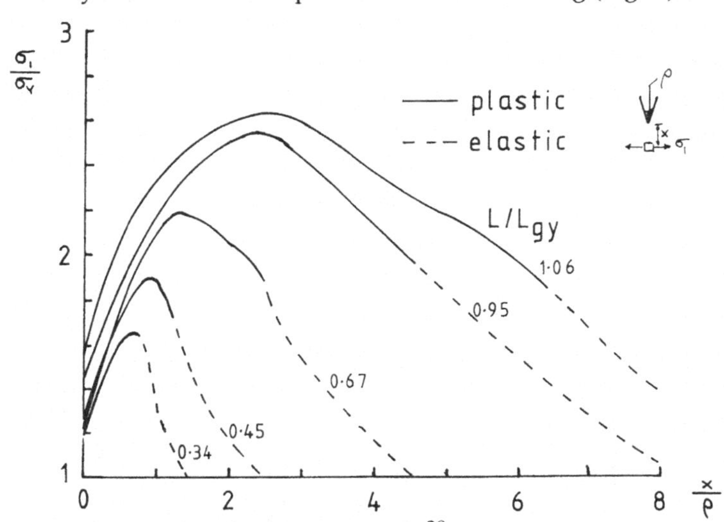

Fig. 1. Plastic stress distributions at a notch.[20]

given fixed position ahead of the crack tip, the tensile stress first increases and then decreases with increased loading as the peak level passes through. The peak shear stress on the other hand occurs at the crack or notch tip position. The maximum effective stress is given by a combination of direct and shear stresses including an allowance for friction and dislocation pile up stresses as shown in equation 1.

$$\sigma_{eff} = \sqrt{\{\sigma^2 + \tau_i^2 \frac{d}{c} \left[1 + \frac{4}{\pi}\left(\frac{c}{d}\right)^{\frac{1}{2}} \frac{\tau_i}{\tau_e}\right]^2\}} \tag{1}$$

where σ, τ are direct and shear stresses
τ_i, τ_e are lattice friction and effective shear stresses
c, d are particle and grain sizes,

$$(\tau_e = \tau - \tau_i)$$

The position of the highest value of effective tensile stress lies between the crack tip and the peak tensile stress position. Gibson[18] showed from careful fractographic measurements that where cleavage fracture had initiated from small second phase particles the initiation position corresponded to that of maximum effective shear stress and was always less than the distance to the peak tensile stress location. Since this initiation position would have previously experienced a higher peak tensile stress as the load was increased he concluded that fracture initiation could not be due solely to maximum tensile stress but must be due to effective tensile stress including the shear stress contributions. Gibson also noted that for large particles (greater than 1·5 μm thick) the nucleation sites were generally located significantly further away from the crack tip than peak tensile stress. He concluded that for small brittle particles the critical controlling phenomenon was propagation of particle sized microcracks into the surrounding ferrite matrix, whereas for larger particles the controlling phenomenon was the cracking of the particle itself, with subsequent propagation into the matrix and across grain boundaries following automatically.

Ductile fracture

Cottrell noted, in conjunction with work by Puttick, that ductile fracture occurred by growth of plastic cavities generally nucleated at inclusions.[4] Quantitative analysis of this phenomenon was carried out by McClintock[21] and by Rice and Tracey.[22] The McClintock analysis took into account strain hardening behaviour whereas the Rice and Tracey analysis was for a non-hardening material, in both cases leading to expressions for the rate of increase of hole size as a function of plastic strain and stress state. This in turn led to the concept of a critical limit to hole growth for ductile fracture expressed by equation 2.

$$\int_1^2 0{\cdot}283 \exp\left(\frac{3\sigma_m}{2\sigma_{eq}}\right) d\epsilon_{eq}^p = \ell n \left(\frac{R}{R_o}\right)c \tag{2}$$

where σ_m, σ_{eq} are mean and effective stresses
ϵ_{eq}^p is effective plastic strain
$(R/R_o)_c$ is critical ratio of cavity growth for fracture.

Quantitative analytical and experimental work on effects of stress state on ductile fracture was reported in the work of Hancock *et al.*[23,24] and Fig. 2 is reproduced from this work. Subsequent work by Beremin[25] has led to analysis methods for ductile fracture which take into account both inclusion size and spacing. This work is being developed in the local approach to fracture, described later.

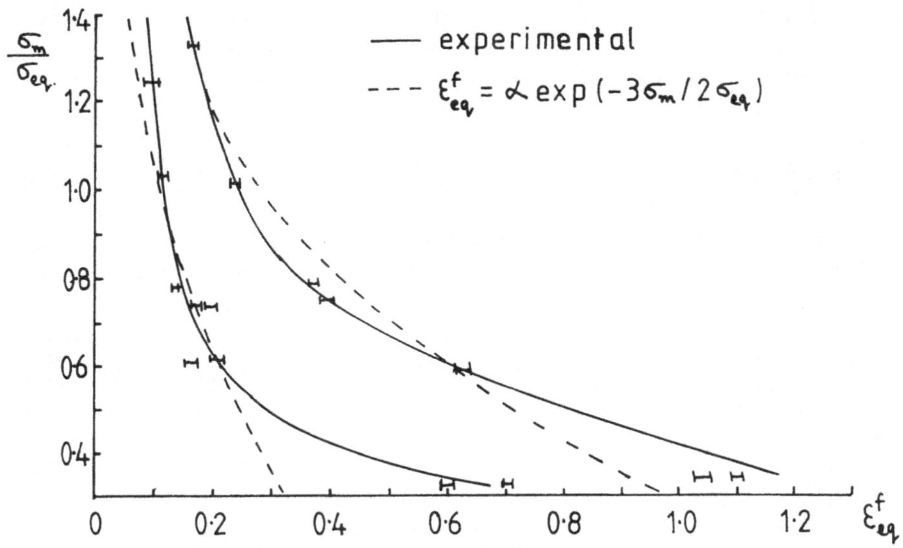

Fig. 2. Effect of triaxiality on ductile fracture.[23]

Applications of Fracture Mechanics to Assessment of Failure

Basic fracture mechanics concepts

The basis of any fracture mechanics analysis is a comparison of applied conditions at the crack tip with the material resistance. The earliest analysis was the classical work of Griffith[26] based on an elastic theory comparison of strain energy released by incremental crack extension with that required to create new fracture surfaces. Irwin[27] and Orowan[28] showed independently that the basic elastic theory analyses would remain valid for materials which failed with local crack tip plasticity provided the surface energy term of the material resistance was replaced by a term representing plastic work. Irwin[29] went on to found the basis of modern fracture mechanics treatments by the introduction of the stress intensity factor *K*, as a crack tip characterising parameter, and showed that for linear elastic conditions *K* was uniquely linked to Griffiths' strain energy parameter. Subsequently Rice[30] extended these analyses to non-linear elastic materials, with the concept of the J-contour integral, which is now the basis of many modern elastic plastic fracture mechanics treatments. Rice introduced the J-contour integral as a measure of strain energy potential, but subsequently Hutchinson[31] and Rice and Rosengren[32] showed that J had a crack tip characterising role for non-linear materials and the stress and strain field was uniquely defined by J for a specific zone around the crack tip (known as the HRR zone).

In order to apply any fracture mechanics treatment it is necessary to have a modelling relationship between external loading conditions, geometry and local crack tip conditions, and it is also necessary to have experimental measurement of the material's resistance by a compatible technique (fracture toughness).

Prior to the introduction of modern powerful elastic plastic finite element stress analysis programs it was necessary to make the best of simplified models. One model which has received extensive consideration, and has made useful insights into elastic plastic behaviour has been the strip yield model, first considered by Dugdale.[33] The analyses by Bilby, Cottrell and Swinden[6,7] although based on dislocation theory were in fact developments of the strip yield model. Similar analyses were carried out by Burdekin and Stone[34] leading to the relationship between crack opening displacement and applied loading conditions for a central crack in a tensile plate of elastic plastic material. This relationship is as follows:

$$\delta = \frac{8\epsilon_Y a}{\pi} \ell n \sec\left(\frac{\pi\sigma}{2\sigma_Y}\right) \tag{3}$$

where δ is the opening displacement at the tips of a crack of length $2a$, in a plate of yield strength σ_Y (strain ϵ_Y) under remote stress σ.

This approach was of great importance in the development of assessment methods in the UK because it led independently to the BSI PD6493 procedure and the CEGB R6 procedure.

The PD6493/R6 procedures

Experimental work on large scale edge notched tensile tests showed that the theoretical relationships of the strip yield model (equation 3) were not observed in practice. This led Burdekin and Dawes[35,36] to develop the COD design curve which became the basis of part of the formalised procedures for assessment of the significance of defects, issued as a draft in 1975 and eventually published as PD6493 in 1980.[37] The design curve relationship is shown as equation 4.

$$\frac{\delta}{2\pi\epsilon_Y a} = \left(\frac{\epsilon}{\epsilon_Y}\right)^2 \qquad \frac{\epsilon}{\epsilon_Y} \leqslant 0{\cdot}5 \tag{4}$$

$$\frac{\delta}{2\pi\epsilon_Y a} = \frac{\epsilon}{\epsilon_Y} - 0{\cdot}25 \qquad \frac{\epsilon}{\epsilon_Y} > 0{\cdot}5$$

This was the first attempt at practical guidance for assessing the significance of defects on a fitness for purpose basis, and it should be noted that the PD6493 document covered fracture under linear elastic as well as yielding conditions, and also covered fatigue and other modes of failure. The document has served well for a number of years and is currently being revised to take advantage of recent developments.

The CEGB R6 procedure originally took as its basis the two criteria approach of Dowling and Townley[38] with interpolation between the limiting conditions of linear elastic fracture and plastic collapse being provided by the strip yield model.[39] Extensive validation of the CEGB R6 approach has been carried out in recent years, particularly in respect of pressure vessel applications for the nuclear industry, and as time has progressed the approach has been refined and modified resulting in the issue of Revision 3 of the procedure in 1986.[40]

The basis of the R6 approach is an assessment diagram in which the ordinate axis represents the ratio of elastic crack tip severity to fracture toughness (the fracture axis K_r) whilst the abscissa axis represents the ratio of applied load to collapse load (the collapse axis S_r or L_r). The current R6 Rev. 3 procedure offers three options of assessment diagram based on the level of detail with which the material properties are known and three categories of analysis which may be based on first initiation of fracture, a limited amount of tearing or a fully detailed tearing analysis. Figure 3 shows the R6 Rev. 3 assessment diagram and assessment is carried out by plotting a point corresponding to the required crack size and loading conditions as shown at point A on the figure. Provided this point lies inside the assessment curve it is deemed safe, whereas if it lies outside the assessment curve it is unsafe. The distance of the point

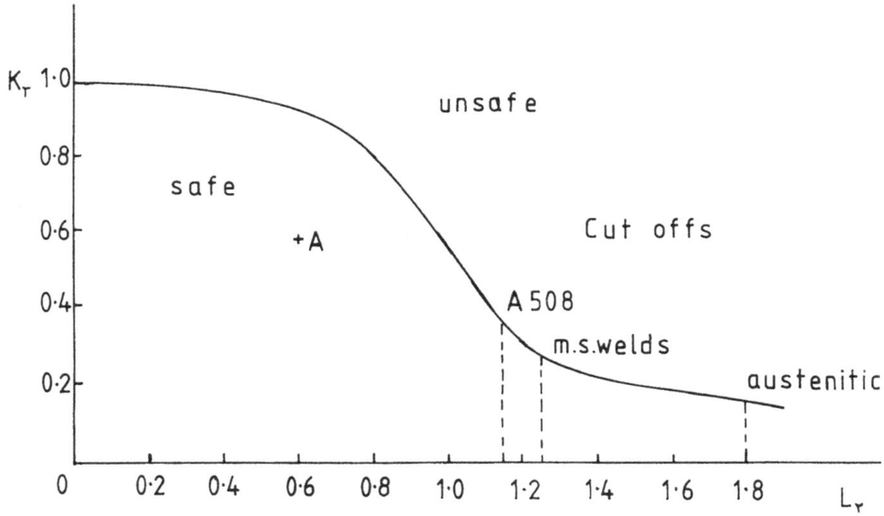

Fig. 3. R6 Rev. 3 assessment diagram.[40]

from the assessment line can be used as an indication of the margin of safety.

It is important to note that in the R6 assessment diagram, the ordinate axis uses only the elastic component of the crack tip severity parameter, and the effects of plasticity in increasing the crack tip severity are allowed for in the reduction in height of the assessment diagram as the plasticity increases across the abscissa. Thus the shaded height of the diagram in Fig. 4 represents the effect of plasticity in reducing the available elastic component of toughness. In principle the assessment diagram would be expected to be dependent on both material stress/strain curve and geometry, but for practical purposes the Rev. 3 assessment curves are based on J-integral analyses of different material groups to represent material stress/strain behaviour and examples of practical geometries from finite element analysis. Whilst extensive analyses have been carried out on geometries relevant to pressure vessels, further checks are desirable for radically different geometries to confirm that the basic shape of the assessment diagram remains safe.

The revisions to PD6493 will attempt to bring together the R6 and earlier PD6493 treatments. This will offer assessments at three levels, an initial screening level compatible with the existing 1980 edition of PD6493, a normal level based largely on CEGB R6 Rev. 2, and an advanced treatment for high strain hardening materials or tearing analyses based on R6 Rev. 3. The assessment diagram approach forces consideration of both plastic collapse and fracture and their interaction, and provides a valuable visual presentation of margins of failure. Complexities in

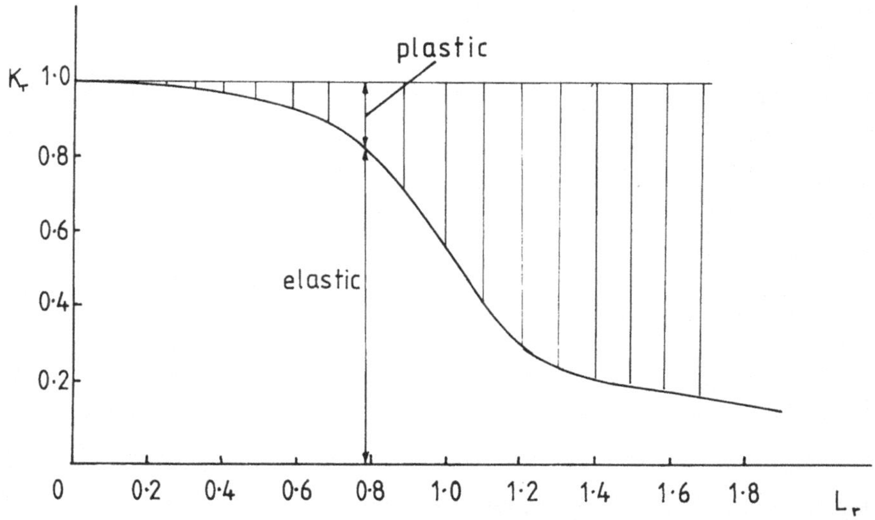

Fig. 4. Assessment diagram allowance for effects of plasticity on crack tip driving force.

carrying out practical assessments satisfactorily remain in the areas of residual stresses and stress concentrations, effects of constraint and thickness on toughness values, and the problems of scatter and reliability.

Tearing analysis treatments are generally based on the concept of a tearing resistance curve which may be expressed in terms of J or COD (Fig. 5). The concept of tearing modulus was introduced by Paris as follows:

$$T = \frac{E}{\sigma_Y^2} \frac{\mathrm{d}J}{\mathrm{d}a} \tag{5}$$

Similar analyses were carried out by Turner[42] and the whole area of tearing analysis and stability of ductile tearing remains one in which research is continuing. This is particularly the case in respect of the extent to which a tearing resistance curve is size or constraint dependent, and its interpretation to tearing in a full structural assessment. A tearing analysis can be carried out using an assessment diagram by finding the conditions which cause an assessment point outside the assessment line to move along a curve which just becomes tangential to the assessment line, the point of criticality being the tangency point (Fig. 6).

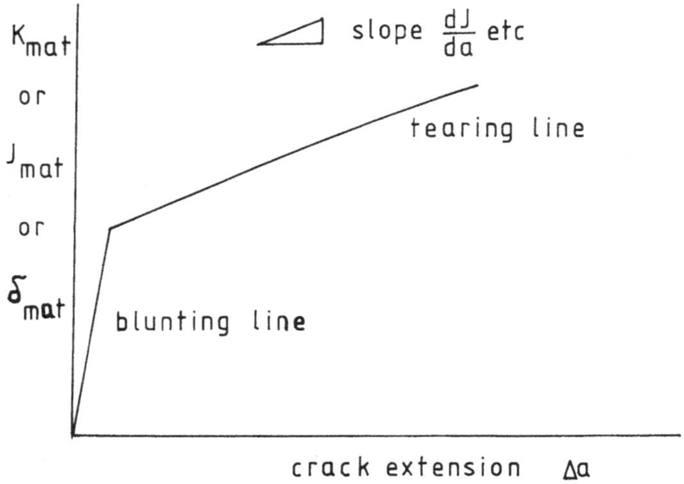

Fig. 5. Schematic ductile tearing resistance curve.

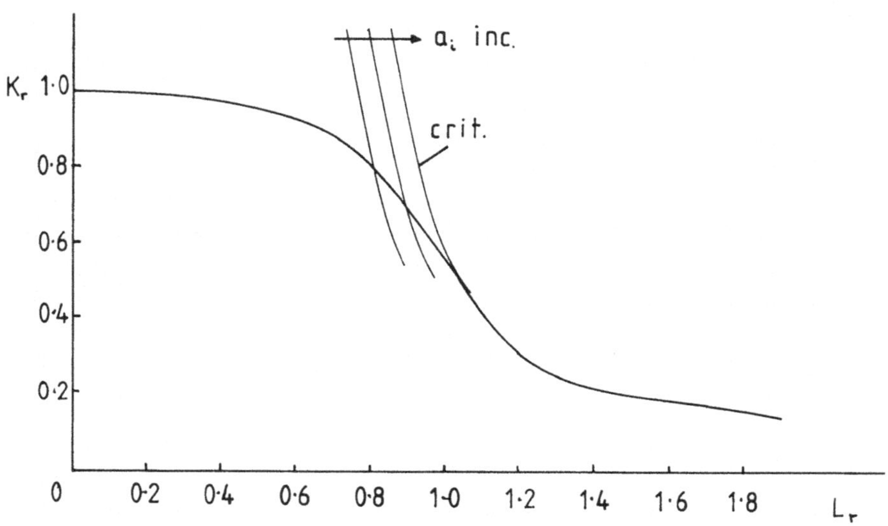

Fig. 6. R6 tearing assessment.

The Marshall/Hirsch LWR Study Group Reports

A Study Group on the integrity of Pressurised Water Reactor pressure vessels was set up under the Chairmanship of Dr. Walter Marshall in 1974 and produced its first report in 1976.[43] The Study Group was reconstituted in 1980 to update its first report and issued its second report in March 1982.[44] An addendum to the second Study Group report was prepared under the Chairmanship of Sir Peter Hirsh to deal with any ongoing matters particularly in the light of the decision to proceed with the construction of the Sizewell 'B' PWR station. The addendum report was published in 1987.[45]

In the first Study Group report the effect of material selection, design, fabrication, and non-destructive testing were considered on the reliability of the reactor vessel of a generic four loop plant design, under normal, upset and test, emergency, and fault conditions. It was concluded that for normal, and upset and test conditions the most probable value of K_{Ic} on the upper shelf was about 240 MPa/m and the critical depth of semi-elliptical crack for these conditions would exceed 100 mm. The probability of failure during normal operation or during upset and test conditions was calculated as less than 10^{-6} per vessel year. Under emergency and fault conditions the first Study Group report calculated critical crack depths for crack extension of not less than 25 mm with a failure probability of less than 10^{-4} per emergency/fault occurrence with such events having a predicted frequency in the range 10^{-4} to 10^{-2} per vessel year.

In the second Study Group report[44] assessments of limiting flaw sizes were carried out using the CEGB R6 Rev. 2 procedures[39] with extensive series of calculations carried out for different conditions. By the time of the second report, the Study Group had arrived at a firm preference for the use of ring forgings of ASTM A508 Class 3 steel, with restrictions on carbon content to reduce risks of heat affected zone cracking during fabrication, and restrictions on sulphur content to less than 0·010% to give high ductile tearing resistance on the upper shelf. Again the assessment was carried out on a generic four loop design vessel, with all nozzles placed in a single belt, and with ring forgings which reduced the number of welded joints, particularly in the belt line irradiated zone. An extensive review of materials toughness data was carried out and the results presented in the form of mean and mean minus two standard deviations curves for static loading and for crack arrest, with the static loading case being extended to consider tearing resistance up to a ductile crack extension of 2 mm for both A533B plate material and A508–3 forging material. The lower bound curves are shown in Fig. 7. A careful series of stress analysis calculations was carried out for each of the loading conditions to be assessed. The linear elastic stress intensity factor calculations were

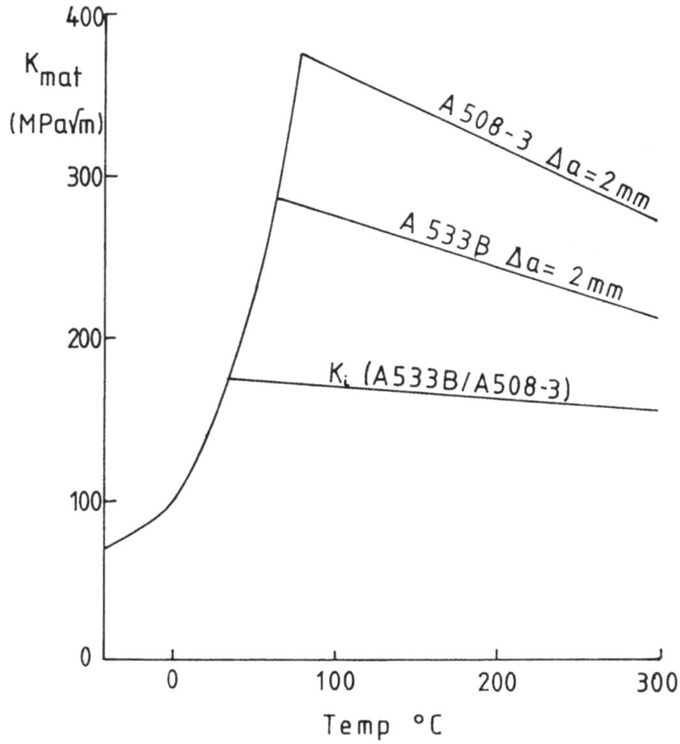

Fig. 7. Lower bound toughness data from second LWR Study Group report.

carried out using validated computer solutions for three crack shapes, namely an extended line crack, a six to one elliptical surface crack, and a semi-circular crack. Fracture mechanics calculations using the R6 Rev. 2 procedures were then carried out to determine the limiting flaw size for the initiation of failure in normal operating, upset and test, and emergency conditions. In all of these cases the minimum limiting flaw size for an extended defect was found to be 60 mm deep, and for a semi-elliptical defect 95 mm deep based on nominal properties, the most severe case being the initial hydrotest. In most other cases the limiting size of semi-elliptical defect was found to be at least 125 mm crack depth. Fault conditions were studied extensively with particular reference to the loss of coolant accident (LOCA) and the pressurised thermal shock (e.g. steam-line break). The thermal stress distributions resulting from such accidents were calculated, and the complex situation assessed of behaviour of cracks at the inner surface of the vessel, chilled by emergency cooling water to give the possibility of high thermal stresses at low temperatures. The worst cases to be considered were a small steam line break in which combined thermal and pressure stresses could occur, and the large loss of coolant accident. For the steam line break the limiting flaw size for initiation was calculated at about 20 mm depth for an extended flaw at lower bound toughness of 170 MPa/m, but with an elliptical flaw of aspect ratio six to one the flaw depth at initiation became 31 mm for the same conditions. If the temperature of the vessel remained such that the minimum lower bound toughness was maintained the limiting flaw sizes for the LOCA were similar to the steam line break, but if the toughness dropped to lower values in the transition region, the calculated limiting flaw sizes became much smaller leading to the possibility of crack extension along the surface leading to crack arrest within the wall thickness at the higher temperature and toughness conditions prevailing there. If the enhanced toughness associated with stable tearing was considered it could be shown that unstable fracture would not occur at flaw sizes less than about 50 mm depth. The conclusion about enhanced toughness for stable tearing was limited by the Study Group's view that stable tearing should not be permitted to exceed a J controlled zone for either the fracture toughness test specimens or the service condition, and this amount of tearing should be limited to 2 mm maximum. The benefits of enhanced toughness from stable tearing clearly rely upon such tearing taking place, and there being no intervention of cleavage fracture. The Study Group was therefore concerned that the possibility of cleavage fracture in the fault condition should be avoided and expressed a preference for this to be achieved by the vessel remaining on the upper shelf, i.e. above the transition temperature range. The second report of the Study Group also reviewed developments and improvements in non-destructive testing, and emphasised the importance of using diverse and redundant techniques capable of detecting flaws well below the fitness for purpose size. The group further recommended that these inspection procedures should be formally validated by an independent organisation.

After the decision was announced to proceed with the construction of the Sizewell 'B' PWR station, the Study Group was reconvened under the Chairmanship of Sir Peter Hirsch to deal with any generic issues which had arisen from the Public Enquiry into the Sizewell 'B' Project. Amongst its other deliberations the Study Group extended further its analyses of stresses during fault conditions to identify the most severe cases. Sensitivity studies were carried out on the effects of thickness of the stainless steel cladding, temperature of the emergency core cooling water supply, and on different severities of thermal shock. The results of the main second Study Group Report were broadly confirmed. It was concluded that the most severe overall fault conditions which could possibly arise would be those during a pressurised thermal shock accident. Fracture mechanics analyses using the R6 Rev. 2 procedures showed that, provided the toughness did not fall below that assumed for the upper shelf the limiting flaw size for the most severe conditions postulated should not fall below 25 mm depth and the probability of failure still estimated as not less than 10^{-7} per vessel year provided there was no correlation between the occurrence of weld defects and regions of poor fracture toughness, and the enhanced toughness associated with stable tearing was permitted for the fault condition analysis.

This series of reports represents an outstanding example of use of fracture mechanics procedures for assessment of both ductile and brittle failure in critical components where the probability of failure must be kept to an incredible level.

The local approach to fracture

Considerable interest has been aroused in recent years by developments in what has become known as the local approach to fracture. Essentially this involves consideration of the detailed stress and strain conditions in small volumes around a crack or stress concentration region by elastic plastic finite element analysis, and the use of failure models for brittle or ductile failure based on small scale material tests usually in the form of notched tensile tests. An International Seminar into developments of the local approach was held in France in 1986, and the published proceedings include a number of papers describing different aspects of the approach.[46–48] In the local approach applied to cleavage fracture Mudry[46] describes the application of Weibull statistics to assess the probability of microcracks becoming unstable in an elemental volume. The probability of microcracks not causing failure at stress σ_1, in volume V, is given by the following expression:

$$p_{nF}(V,\sigma_1) = \exp\left[-\left(\frac{\sigma_1 - \sigma_{th}}{\sigma_u}\right)^m \cdot V\right] \tag{6}$$

where σ_{th} is a threshold stress level, σ_u is the critical fracture stress, and m is the Weibull modulus. The detailed stress field is computed by finite element stress analysis with the fracture of each small volume predicted from the history of stresses and strains in that volume, and the total effect summed up over all elements in the plastic zone region. Behind this lies the assumption that microcracks are created as soon as plasticity begins. The mesh size is chosen to suit stress gradients so that elemental volumes can be treated as being under uniform stress conditions. Analysis of this kind is carried out on a series of small notched tensile tests with different notch root radii to determine an effective Weibull stress from the following relationship:

$$\sigma_w^m = \sum \sigma_1^m \frac{\Delta V}{V_u} \tag{7}$$

Enough tensile tests are carried out to determine both the Weibull stress and scatter parameters which are consistent with the finite element analysis. This type of approach can be used to predict K_{Ic} fracture toughness values from the following expression:

$$\sigma^m = \frac{K_{Ic} \cdot B \cdot \sigma_Y^{m-4} \cdot C_m}{V_u} \tag{8}$$

and the scatter in K_{Ic} from equation 9:

$$p_{nF}\,(K_{Ic}) = \exp\left[-\left(\frac{\sigma_w}{\sigma_u}\right)^m\right] \tag{9}$$

where this applies for low hardening materials, and C_m is a non-dimensional constant with a value of $1{\cdot}5 \times 10^{-6}$ for a value of the Weibull exponent m of 22. The parameter V_u is a characteristic volume which is the inverse of N_v the number of microcracks per unit volume, typically taken as 5 to 10 per volume of grain size.

For application to practical problems, σ_u, m, and V_u are determined for the particular material from a series of small scale tensile tests. The finite element analysis of the practical problem requires a good simulation of large strain plasticity, special treatments for incompressibility in plasticity, and a satisfactory representation of crack blunting. For rigorous accuracy the mesh size should be much smaller than the crack opening displacement at fracture, but Mudry concludes that an adequate mesh size is related to the characteristic volume V_u for the low toughness end of material results.

Figure 8 shows the use of the local approach to calibrate compact tension and centre cracked tension test plates to give the minimum ligament sizes for which K_{Ic} values can be satisfactorily determined from J tests. This cleavage fracture approach has been applied to practical problems of thermal shock in cylinders, to the behaviour of underclad defects, and to mixed mode stress conditions. It is claimed that better predictions are achieved than by using linear elastic K_{Ic} results, but the main reason for this appears to be that a lower toughness is predicted for a long crack front than would be obtained from standard K_{Ic} tests.

The application of the local approach to ductile fractures[47, 48] again uses mechanistic models for the failure mode, in this case microvoid coalescence for ductile fracture, calibrated by simple and inexpensive notched tensile tests such as those of Hancock *et al.*[23], as shown in Fig. 2. The finite element stress analysis is now used to predict hole growth

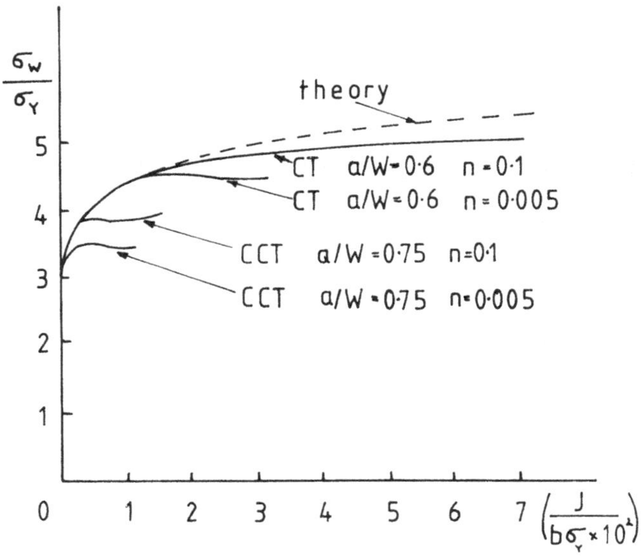

Fig. 8. Limits on specimen geometry for converting J_{Ic} *to* K_{Ic} *results.*

within elemental volumes based on inclusion size and spacing and the overall effect integrated throughout the plastic zone volume.

The big advantage offered by the local approach is the potential for deriving fracture toughness test results relevant to particular constraint situations from small scale simple inexpensive tests, including determination of anticipated scatter, without having to do large numbers of large and expensive tests. The disadvantages are the extensive finite element analysis required, and the expense of this for three-dimensional elastic plastic situations. Nevertheless with improved understanding of the modelling of mechanisms of fracture, the advances in computational and stress analysis capability should lead to improved characterisation of both ductile and brittle failure in practical situations.

Developments in Non-Destructive Testing

Just as there have been substantial improvements in understanding of mechanisms of fracture, and the application of failure assessment methods, significant improvements in capability and reliability of non-destructive testing as called for in the Marshall/Hirsch reports[43–45] and by Cottrell[1] have been achieved. Substantial improvements have taken place in ultrasonic testing both in respect of detection of flaws and in ability to size flaws. The PISC II[49] and defect detection trials[50] in which several groups were invited to develop non-destructive testing procedures and inspect blocks containing artificial flaws without knowledge of these flaws, have shown that the best techniques are capable of detecting flaws of 6 mm size with high reliability, and flaws of 20–25 mm size with extremely high consistency and reliability. Sizing accuracy has also been improved in the best of these techniques, using automatic scanning procedures with multiple angles and computer data storage and processing. With time of flight techniques, sizing accuracies of the order of ±3 mm or better can be obtained, and even with conventional pulse echo techniques it is possible to establish a clear margin between fitness for purpose flaw sizes, and reliable detection limits.

Whilst the development of these improvements in ultrasonic testing capability has resulted from basic research assessing signal response from different sizes, types and roughnesses of flaw, a significant contribution has also been made by the establishment of the Inspection and Validation Centre at UKAEA Risley. The purpose of this independent Centre is to ensure that all ultrasonic testing procedures and operators are capable of detecting and sizing any flaws which may be of consequence in the reactor pressure vessel of the PWR system by both theoretical and experimental assessment of capabilities using test blocks containing a wide range of different sizes and types of flaw. The results of validation trials have been very successful, confirming that both procedures and operators used on the inspection of practical construction of PWR vessels, are capable of achieving reliable detection of any flaws which might be of concern on a fitness for purpose basis. The validation trials have shown significantly better capability than this with a high probability that much smaller flaws can be detected. It is significant that the Chairman of the Management Advisory Committee of the Inspection and Validation Centre is Sir Alan Cottrell.

Conclusions

The pioneering work of Sir Alan Cottrell in the field of fracture has led to substantial developments in understanding the basic mechanisms of both brittle and ductile fracture in steels.

Developments in failure assessment methods have now taken place to the extent that safe and reliable assessments can be made of the significance of defects on failure in critical construction. Significant improvements have also taken place in non-destructive testing, such that the threshold for reliable detection of defects is well below defect sizes which would be of concern in modern high quality construction fabricated from modern tough steels. These developments give great confidence that safe and reliable construction can be achieved in critical structural components.

References

1. A. H. Cottrell, 'How safe is nuclear energy', Heinemann, London, Exeter N.H., 1981.
2. A. H. Cottrell, UKAEA Conference on Brittleness of Metals, Conf. Brit./P.P.I., 1957.
3. A. H. Cottrell, *Trans. A.I.M.E.*, **212**, 192, 1958.
4. A. H. Cottrell, 'Theoretical aspects of fracture,' in Fracture Proceedings of Swampscott Conference (ed. Averbach *et al.*). M.I.T. and John Wiley, N.Y., Chapman and Hall Ltd., London, pp. 20–44, 1959.
5. A. H. Cottrell, I.S.I. Special Report No. 69, p. 281, 1961.
6. B. A. Bilby, A. H. Cottrell and K. H. Swinden, *Proc. Roy. Soc. Ser.* **19**, Vol. 272, 1963.
7. B. A. Bilby, A. H. Cottrell, E. Smith and K. H. Swinden, *Proc. Roy. Soc. A*, Vol. 279, 1964.
8. A. H. Cottrell, The Bakerian Lecture 1963, *Proc. Roy. Soc. Ser.* **19**, Vol. 276, pp. 1–18, 1963.
9. J. F. Knott and A. H. Cottrell, 'Notch brittleness in mild steel', *J.I.S.I.*, **201**, 249–260, 1963.
10. N. J. Petch, UKAEA Conference on Brittleness of Metals, Conf. Brit./P.P.2, 1957.
11. N. J. Petch, *Ibid*, ref. 4, pp. 54–64.
12. A. A. Wells, Crack Propagation Symposium, College of Aeronautics and Royal Aeronautical Society, Cranfield, 1961.
13. J. W. Davies and A. A. Wells, *Nature*, **190**, 432, 1961.
14. E. Smith, Conference on Physical basis of yield and fracture, Inst. Phys. and Phys. Soc., London, pp. 36–46, 1966.
15. E. Smith and J. T. Barnby, *Met. Sci.*, **1**, 56–64, 1967.
16. J. F. Knott, *J.I.S.I.*, **204**, 104, February 1966.
17. R. O. Ritchie, J. F. Knott and J. R. Rice, *J. Mech. Phys. Sol.*, **21**, 395–400, 1973.
18. G. P. Gibson, 'Review of cleavage fracture mechanisms in ferritic steels', UKAEA Report R13227, Harwell, Oxon, 1988.
19. R. M. McMeeking, *J. Mech. Phys. Sol.*, **25**, 357–381, 1977.
20. J. R. Griffiths and D. R. J. Owen, *J. Mech. Phys. Sol.*, **19**, 419–431, 1971.
21. F. A. McClintock, 'A criterion for ductile fracture by the growth of holes', *J. Appl. Mech.*, pp. 363–371, June 1968.
22. J. R. Rice and D. M. Tracey, 'On the ductile enlargement of voids in triaxial stress fields', *J. Mech. Phys. Sol.*, **17**, 201–217, 1969.
23. J. W. Hancock and A. C. Mackenzie, *J. Mech. Phys. Sol.*, **24**, 147–169, 1976.
24. J. W. Hancock and D. K. Brown, *J. Mech. Phys. Sol.*, **31**, 1–24, 1983.
25. F. M. Beremin, 'Experimental and numerical study of the different stages in ductile rupture: application to crack initiation and stable crack growth', in

Three Dimensional Constitutive Relations and Ductile Fracture (ed. Nemat-Nasser). N. Holland, Amsterdam, pp. 185–205, 1981.
26. A. A. Griffith, *Phil. Trans. Roy. Soc. A*, **221**, 163, 1921.
27. G. R. Irwin, Symposium American Soc. Metals, Chicago, p. 147, 1947.
28. E. Orowan, Proc. Symp. Fatigue and Fracture of Metals, New York, M.I.T. and John Wiley, p. 139, 1950.
29. G. R. Irwin, *J. Appl. Phys.*, **24**, 361, 1957.
30. J. R. Rice, *J. Appl. Mech.*, **35**, 379, 1968.
31. J. W. Hutchinson, *J. Mech. Phys. Sol.*, **16**, 337, 1968.
32. J. R. Rice and G. F. Rosengren, *Ibid*, **16**, 1, 1968.
33. D. S. Dugdale, *J. Mech. Phys. Sol.*, **8**, No. 2, p. 100, 1960.
34. F. M. Burdekin and D. E. W. Stone, *J. Strain Anal.*, **1**, No. 2, January 1966.
35. F. M. Burdekin and M. G. Dawes, Conf. on Practical Application of Fracture Mechanics to Pressure Vessel Design, Paper C5/71, I.Mech.E., London, 1971.
36. M. G. Dawes, 'Fracture control in high strength steel weldments', *Weld. J. Res. Suppl.*, **53**, 369s, 1974.
37. BSI Published Document PD6493, 'Guidance on some methods for the derivation of acceptance levels for defects in fusion welds on a fitness for purpose basis', 1980.
38. A. R. Dowling and C. H. A. Townley, *Int. J. Press. Vess. and Piping*, **3**, 77, 1975.
39. R. P. Harrison, K. Loosemore and I. Milne, 'Assessment of the integrity of structures containing defects', CEGB Report, R/H/R6, 1976, Revision 1 — 1977, Revision 2 — 1980.
40. I. Milne, R. A. Ainsworth, A. R. Dowling and A. T. Stewart, CEGB Report R/H/R6 Revision 3 — 1986.
41. J. W. Hutchinson and P. C. Paris, 'Stability analysis of J-controlled crack growth', ASTM Symposium on elastic-plastic fracture, ASTM–STP668, 1977.
42. C. E. Turner, 'Methods of post yield fracture safety assessment', in *Post Yield Fracture Mechanics* (ed. D. G. H. Latzko). Applied Science Pub. Ltd., London, 1979.
43. First Report of LWR Study Group, Chairman W. Marshall, 'An Assessment of the Integrity of PWR Pressure Vessels', HMSO, London, 1976.
44. Second Report of LWR Study Group, Chairman W. Marshall, UKAEA, London 1982.
45. Addendum to Second Report of LWR Study Group, Chairman P. B. Hirsch, UKAEA, London, April 1987.
46. F. Mudry, *J. Nucl. Eng. and Design*, **105**, 65–76, 1988.
47. G. Rousselier, *Ibid*, pp. 97–111, 1988.
48. R. Batisse, M. Bethmont, G. Devesa and G. Rousselier, *Ibid*, pp. 113–120, 1988.
49. S. Crutzen, PISC II Report, CSNI 110, 5th June 1985, and NDT Int., Vol. 18, pp. 243–249, 1985.
50. B. Watkins, *et al.*, *Brit. J. NDT*, **25**, 179–192, 1983; **26**, 87–105, 1984.

4 Dislocation Reactions

P. B. Hirsch and M. J. Whelan†*

*Isaac Wolfson Professor of Metallurgy, University of Oxford
†Reader, Department of Metallurgy and Science of Materials, University of Oxford

Abstract
The interactions of dislocations in face-centred cubic metals and alloys observed by transmission electron microscopy are reviewed first from an historical perspective and then in the light of more recent observations. The central role of Cottrell's mechanism of dissociation of a Lomer edge dislocation on $\{100\}$ to form the Lomer–Cottrell barrier is emphasised.

Introduction — An Historical Aspect

When one of the authors (MJW) returned to the Cavendish Laboratory after the Long Vacation of 1954, he joined the other author (PBH) as a research student in the Crystallographic Laboratory of the Cavendish. The latter suggested that a suitable fruitful field of research might be to try to image dislocations directly in thin crystals with the transmission electron microscope. In fact some seminal work leading up to this had been initiated by A. Kelly[1] in 1953, working in the Cavendish Laboratory. He had collaborated with J. W. Menter of the Laboratory of the Physics and Chemistry of Rubbing Solids (PCRS), using the EM3 electron microscope in PCRS, which was equipped with selected area electron diffraction facilities, to take transmission electron diffraction patterns from beaten gold foils. The work[2,3] was an attempt to obtain small area diffraction information using electrons in transmission, rather than X-rays in back reflection which was a current line of research at the Crystallographic Laboratory of the Cavendish. As a by-product some electron micrographs of beaten gold foil were obtained which showed dark bands of contrast roughly parallel to the traces of $\{111\}$ planes on the foil surface. PBH had taken an interest in this work and it occurred to him that the dark bands might be due to stacking faults, the contrast arising from the phase difference $2\pi\mathbf{g}\cdot\mathbf{R}$ which occurs between waves diffracted above and be-

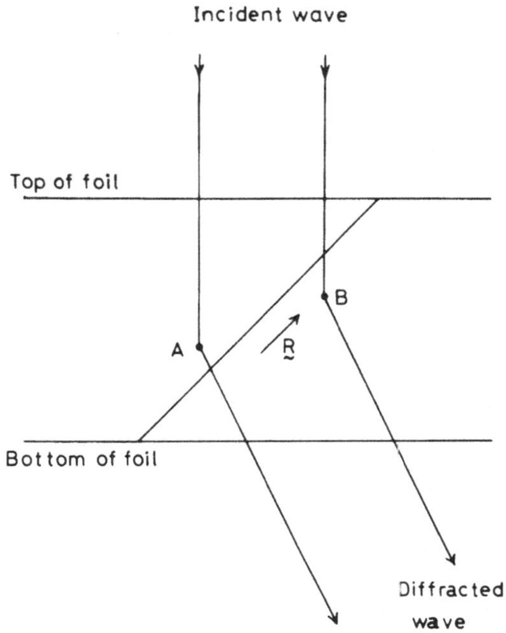

Fig. 1. Cross-section through a foil containing an inclined stacking fault (oblique line). Waves scattered by equivalent atoms A and B on opposite sides of the fault differ in phase by $2\pi\mathbf{g}\cdot\mathbf{R}$ (Whelan[6]).
Courtesy of the Journal of Electron Microscopy Technique.

low the fault plane (Fig. 1) (**g** is the reciprocal lattice vector of the Bragg reflection and **R** is the fault displacement vector). Kelly and Menter had already left Cambridge when MJW started as a research student. PBH showed him the micrographs of gold foil and discussed the contrast mechanism for a fault. Furthermore, because a dislocation on $\{111\}$ in the fcc structure was expected to dissociate into two Shockley partials separated by a ribbon of stacking fault, PBH had a hunch that a dislocation might be made visible due to the ribbon of fault and the above-mentioned phase effect. MJW was unfamiliar with even the concept of a dislocation, let alone partials, since the Physics Tripos which he had just taken contained virtually nothing on solid state defects at that time. His first purchase was therefore a copy of Alan Cottrell's book *Dislocations and Plastic Flow in Crystals*, which cost him 25 shillings, and which he still has on his bookshelf.

Most metallurgical work with the EM, with a few notable exceptions, was at that time concentrated on surface observations by replica techniques. Metallurgists did not think it would be possible to see a dislocation directly in the electron microscope, since it was thought that contrast would arise only from density variation, and this would be too small to cause any effect. When eventually dislocations were observed in aluminium foils,[4] good contrast was observed even though the ribbon-width of stacking fault was virtually negligible owing to the high stacking fault energy. It became clear that contrast arose mainly from the continuous displacement field of an undissociated dislocation, and the fault ribbon mechanism of visibility was almost forgotten. It was many years later that the mechanism was nicely demonstrated by Ray and Cockayne[5] using the weak-beam technique for a dislocation in silicon. Figure 2(a) shows the strong-beam image of an edge dislocation in silicon viewed normal to the

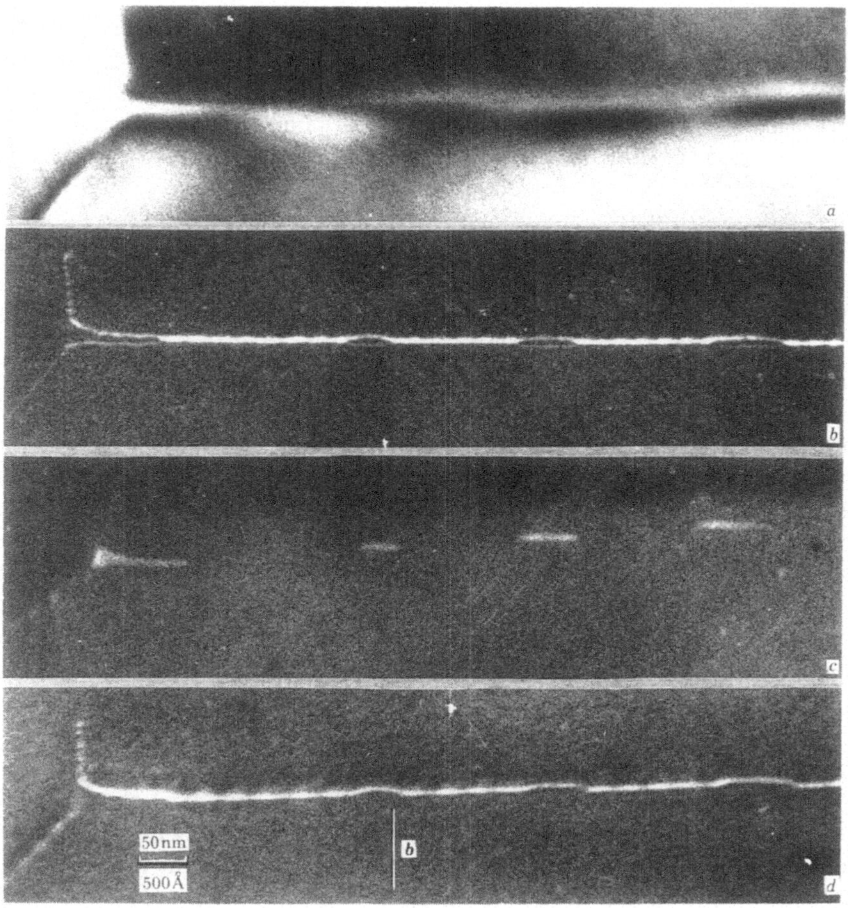

Fig. 2. (a) Strong-beam and (b)–(d) weak-beam images of an edge dislocation in silicon (Ray and Cockayne[5]).
Courtesy of the Royal Society.

slip plane, and no fine detail is visible. In the weak-beam images (Figs.2(b), (c) and (d)) fine detail is visible. Fig. 2(b) uses a {220} reflection for which $\mathbf{g}\cdot\mathbf{b} = 2$ for the whole dislocation, and both undissociated and dissociated segments are visible. Fig. 2(d) uses a {220} reflection in which one of the partials is invisible. Fig. 2(c) uses a {111} reflection for which $\mathbf{g}\cdot\mathbf{b} = 0$ for the whole dislocation and the undissociated segments are invisible. However, the dissociated segments are visible by the fault contrast at the ribbon of fault as in the original idea of 1953–54. A more detailed account of the history of the observation of dislocations by transmission electron microscopy has been given by Whelan.[6]

Early Observations of Dislocation Interactions

In the 1950s several theories of the linear work hardening in stage II of the stress–strain curve of fcc metal single crystals were published,[7–9] which postulated the existence of piled-up groups of dislocations behind barriers in the slip plane. A suitable barrier was thought to be the Lomer–Cottrell dislocation.[10,11] If barriers are formed with primary dislocations of one Burgers vector, the shape of a barrier in the slip plane may be a parallelogram with edges along the two $\langle 110 \rangle$ directions not containing the primary Burgers vector. Theory even suggested the number of dislocations per pile-up to be about 25.[9]

The first observations were made on deformed polycrystalline foils of austenitic stainless steel,[12, 13] and while pile-ups were observed at grain boundaries,[12] the main feature of the dislocation distribution inside the grains was one of tangled networks in which several types of interaction could be identified.[13] Subsequent work on single crystals of pure metals,[14,15] also revealed complex networks and cellular structures, but not the idealized pile-ups.

We now review some of the observations and inferences made.[13] The basic observation was that interactions usually occur over short lengths when intersecting dislocations tangle with a group of dislocations in the primary plane. Figures 3 and 4 illustrate two important cases. In Fig. 3 the primary dislocations with Burgers vector DC on plane (a) (Thompson's notation) are intersected by a dislocation CB on plane (d). A short length of resultant dislocation DB (=DC + CB), shown by a heavy line, is formed at the line of intersection of the slip planes (Fig.3(b)). In this case the intersecting dislocation can cross-slip to the primary plane, so that all three dislocations (primary, intersecting and resultant) can glide on the primary plane to balance the line tension forces at the three-fold nodes. The network shown in Fig. 3(c) is thereby formed. Figure 4 shows the case where the short segments of resultant dislocation AD are Lomer–Cottrell

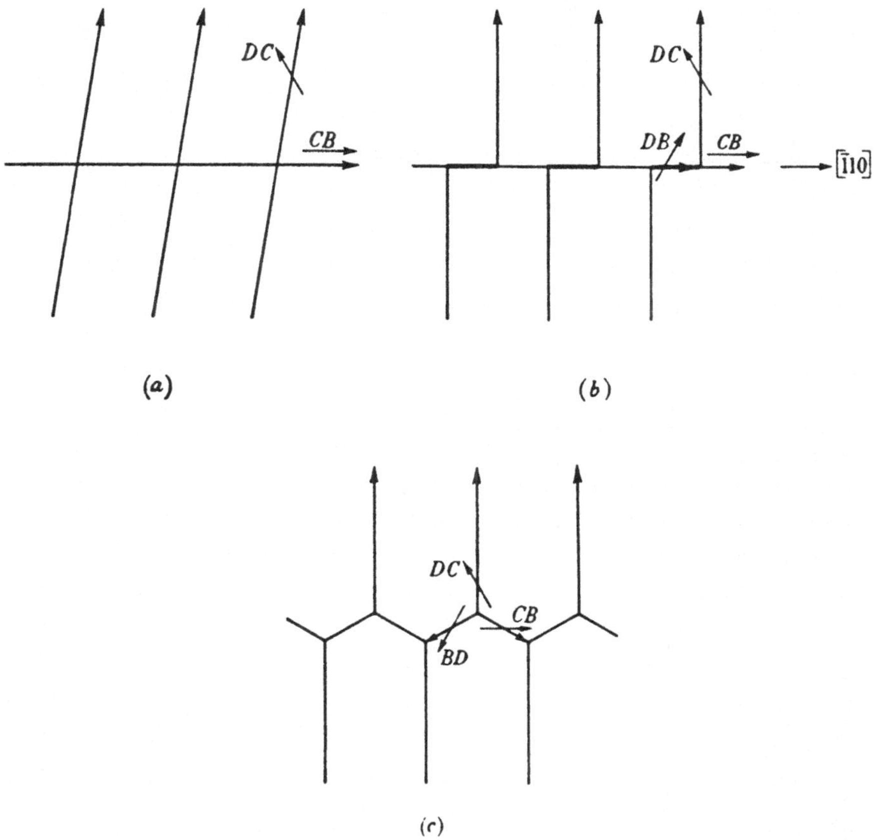

Fig. 3. (a), (b), (c). Interaction of dislocations with Burgers vectors at 120°. (a): A pile-up in the plane of the paper is intersected by a screw. (b): The pile-up dislocations react with the screw to give short lengths of resultant dislocation DB. (c): The arrangement into which (b) degenerates (Whelan[13]).
Courtesy of the Royal Society.

dislocations (LC) with Burgers vector AD (=AC + CD) perpendicular to the line direction CB. This dislocation in its undissociated form can in theory glide on the (001) plane containing AD and CB. Electron micrographs of dislocation interactions thought to represent these cases are shown in Figs. 5, 6 and 7. The network in Fig. 5 is formed by the intersection of a primary group of dislocations by four secondary dislocations as in Fig. 3. Short segments of Lomer–Cottrell dislocation (LC) are visible in Fig. 6 and also in Fig. 7.

To explain the fine detail of the micrographs the interactions must be studied in terms of the constituent dissociated dislocations. Friedel[8] considered this in detail for a two-dimensional model of the interaction, but

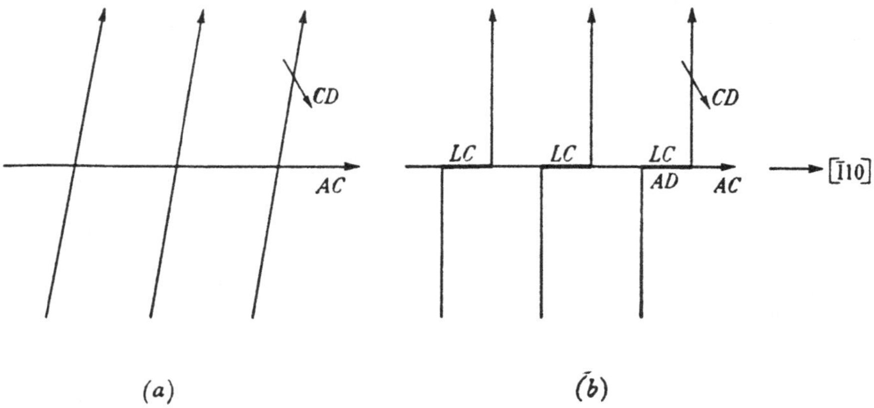

Fig. 4. (a), (b). Interaction of dislocations with Burgers vectors at 120°. Segments of Lomer–Cottrell dislocation (LC) with Burgers vector AD are formed along the line of intersection of the slip plane. The arrangement in (b) cannot degenerate as in Fig. 3(c) because LC is not glissile in the plane of the paper (Whelan[13]).

Courtesy of the Royal Society.

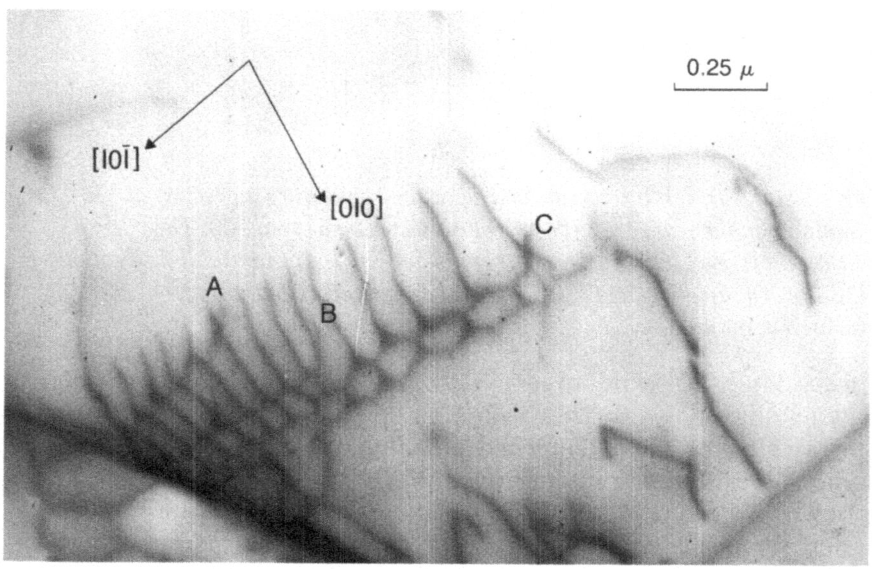

Fig. 5. Hexagonal network formed at a pile-up at a grain boundary in stainless steel. Note the extended and contracted threefold nodes as in Fig. 8(c) (Whelan[13]).

Courtesy of the Royal Society.

the micrographs show that the interaction must be considered as three-dimensional. Figure 8 corresponds to Fig. 3 and shows the interaction in terms of ribbons of intrinsic stacking fault. The primary dislocation (Dα,αC) is intersected by (Cδ,δB) (Fig. 8(a)) and the following reaction occurs.

$$\underset{primary}{(D\alpha + \alpha C)} + \underset{secondary}{(C\delta + \delta B)} \rightarrow \underset{resultant}{(D\alpha + \alpha B)}$$

αB glides on plane (a) and pulls down under the tension force of the stacking fault (Fig. 8(b)). The node indicated with an arrow in Fig. 8(b) is driven along the line of intersection by the line tension of αB and αC until the segment Cδ δB on plane (d) is annihilated. The resultant consists of a network of extended and contracted nodes on the primary plane as shown in Fig. 8(c). The extended and contracted nodes are clearly visible in Fig. 5. Measurement of the shape of an extended node has provided a convenient method of estimating the stacking fault energy.[13, 16]

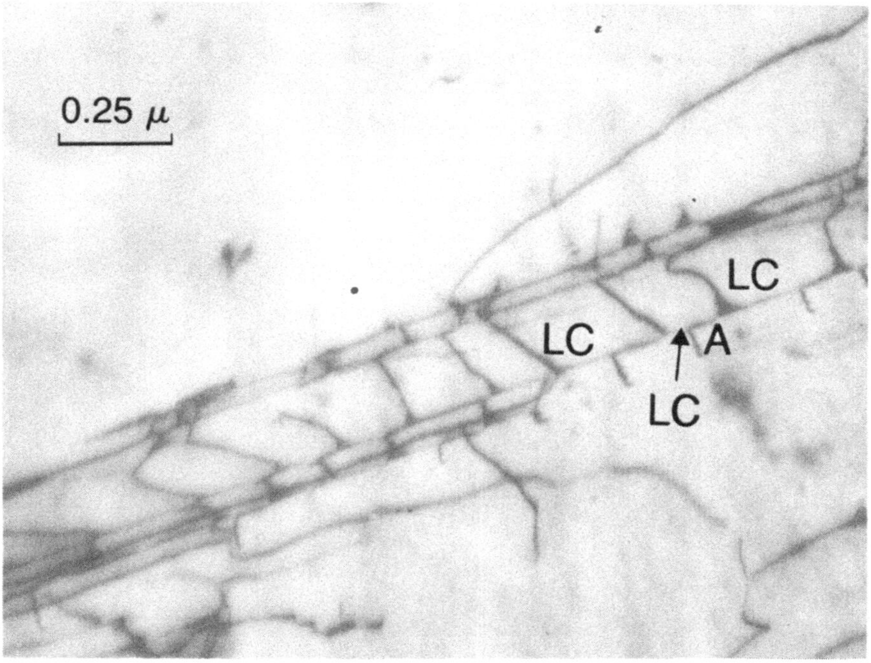

Fig. 6. A number of short primary dislocations has been intersected by several long ones to produce short lengths of resultant dislocation LC, which are probably Lomer–Cottrell locks. Note the extended and contracted nature of the nodes, as illustrated in Fig. 9 (Whelan[13]).

Courtesy of the Royal Society.

(a)

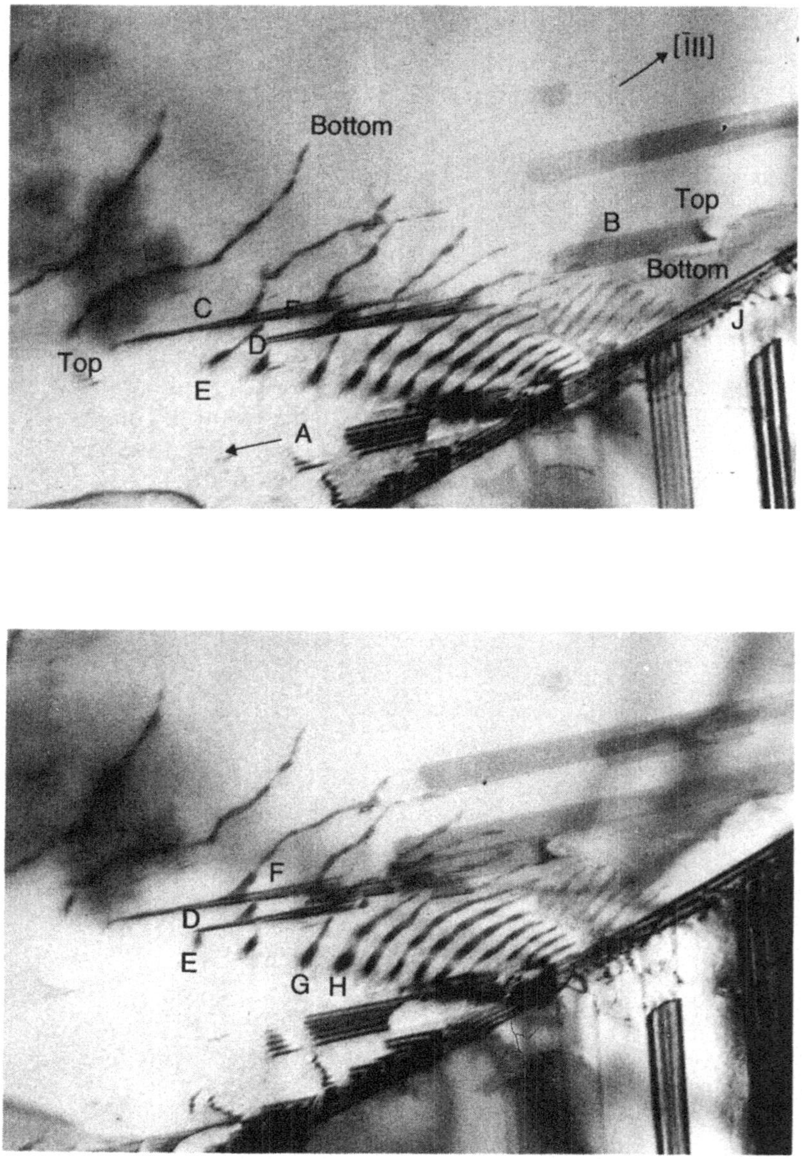

(b)

Fig. 7. (a), (b). Example of Lomer–Cottrell lock formation in situ *during observation. The length EF of primary dislocation in (a) reacts with a partial at D to give the Lomer–Cottrell lock configuration EDF in (b), as illustrated in Fig. 9. Note the extended and contracted nodes at either end of the lock as in Fig. 9 (Whelan[13]).*

Courtesy of the Royal Society.

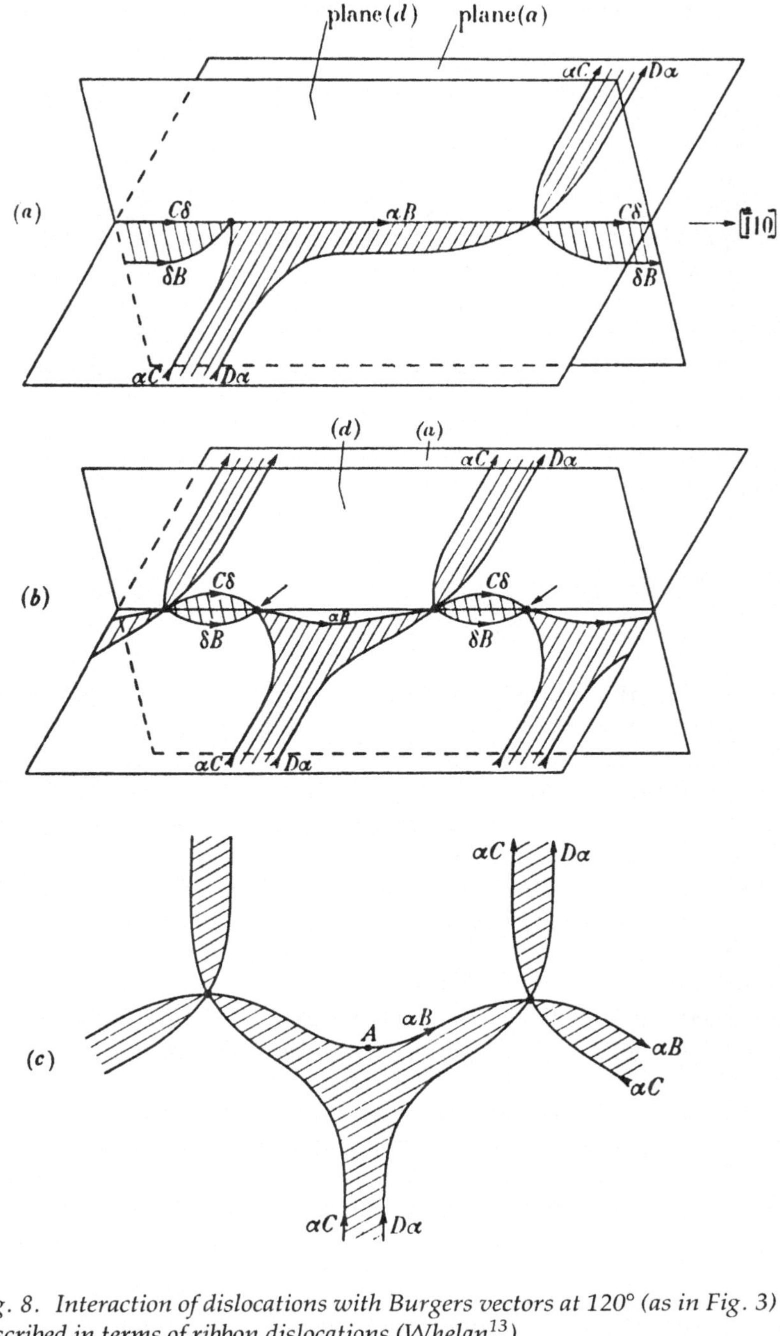

Fig. 8. Interaction of dislocations with Burgers vectors at 120° (as in Fig. 3) described in terms of ribbon dislocations (Whelan[13]).
Courtesy of the Royal Society.

For the case of Fig. 4 the interaction in terms of ribbon dislocations is shown in Fig. 9. The interaction is

$$\underset{primary}{(C\alpha + \alpha D)} + \underset{secondary}{(A\delta + \delta C)} \rightarrow \underset{resultant}{(A\delta + \delta\alpha + \alpha D)} \cdot$$

$\delta\alpha$ is a low-energy stair-rod dislocation, with Burgers vector $\frac{1}{6}\langle 110\rangle$, lying at the acute bend in the stacking fault, while αD and $A\delta$ are Shockley partials glissile on planes (a) and (d). This is the dissociated structure of Lomer's dislocation[10] originally proposed by Cottrell.[11] It is expected that the primary dislocation will have an extended node at one end of the barrier and a contracted node at the other end as sketched in Fig. 9. Extended and contracted nodes are visible in Figs. 6 and 7. Fig. 7(a) and (b) actually shows the Lomer–Cottrell barrier form by the interaction of a partial at D with a segment EF of primary dislocation. These and other possible three-dimensional dislocation interactions were first considered by Whelan.[13]

Further Developments

Theoretical work

Hirth[17] studied the energetics of barrier formation and in a later paper[18] more detailed investigations were made including the effect of anisotropy and asymmetric barriers. Korner *et al.*[19] also considered asymmetric barriers. A notation adopted by Jossang *et al.*[18] is now generally employed in the literature. There are four possible barriers with a stair-rod dislocation with a Burgers vector magnitude $b < \frac{1}{3}|\langle 110\rangle|(2|\delta\alpha|)$. These are denoted by B1 to B4. B1 is the Lomer–Cottrell barrier of Fig. 9. *B*4 is an inverted Lomer–Cottrell barrier, where the other pair of partials of the ribbons ($A\delta$ and αD in Fig. 9) combine to form a stair-rod dislocation with $\mathbf{b} = \frac{1}{3}\langle 110\rangle$. B3 is the so-called Hirth lock (HL) formed by short range interaction of ribbon dislocations with total Burgers vectors at 90° and with a $\frac{1}{3}\langle 100\rangle$ stair-rod dislocation at an obtuse bend in the stacking fault. B2 is another short-range barrier formed by dislocations with total Burgers vectors at 90° and with a $\frac{1}{3}\langle 110\rangle$ stair-rod dislocation at an acute bend in the stacking fault. Details of three of these four (including the Hirth lock) had been given previously, while the fourth (*B*4, inverted Lomer–Cottrell lock) had been considered but discounted by rough energetic reasoning.[8,13]

More recent observations of interaction configurations

The original electron microscope observations of the geometrical features of dislocation interactions[13] had, of necessity, to be rather speculative. At

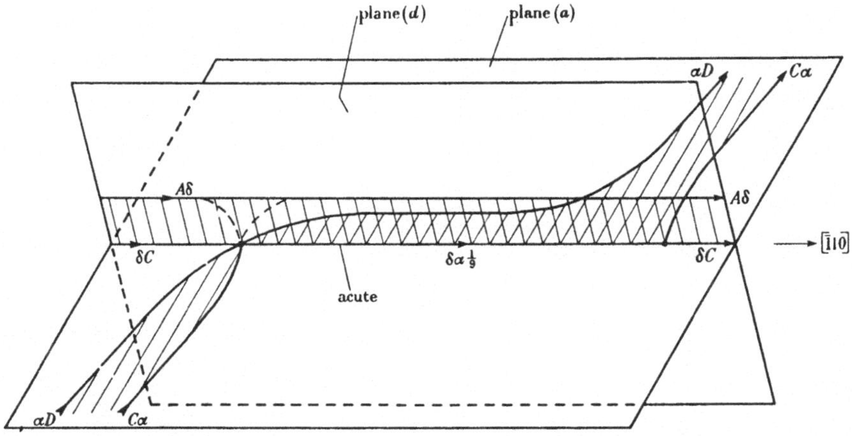

Fig. 9. Interaction of ribbon dislocations to form a Lomer–Cottrell lock (Whelan[13]).
Courtesy of the Royal Society.

that time, 30 years ago, transmission electron microscopy of defects was in its infancy and practical methods of determining Burgers vectors had yet to be developed. Furthermore, the weak-beam technique[20] of revealing dislocation lines at higher resolution had not been developed. Karnthaler and Wintner[21] have rectified these deficiencies by a systematic study of dislocation interactions in Al–Cu alloys of low stacking fault energy (see also Karnthaler *et al.*[22]). The geometry was revealed by high-resolution weak-beam methods, and the Burgers vectors of the dislocations were determined by diffraction contrast techniques. The results confirmed the original predictions.[13] It was found that for single crystals deformed into stage II, interactions occur over short lengths only at the intersections of primary and secondary dislocations, and that most of the resultant dislocations are Lomer–Cottrell locks. Extended and contracted nodes were observed (as illustrated in Figs. 6, 7 and 9). An example is shown in Fig. 10, where the line diagram is a sketch of the interacting dislocations with all Burgers vectors labelled. An interesting observation is the conversion of what would have been the contracted node (here uppermost) into an extended node bounding an inverted Lomer–Cottrell lock (B4), as indicated at the bottom of the diagram. To the left of A′ a segment of B4 has reverted to the normal Lomer–Cottrell LC. At the other lock near A″ no reversion has occurred. The intersecting dislocation S_1, bowed-out on the intersecting slip plane is clearly visible. Hirth locks were very rarely observed, and B2 locks were not observed at all. This might be expected on simple energetic grounds, since taking account of the effect of dislocation character on line energy, one would expect only

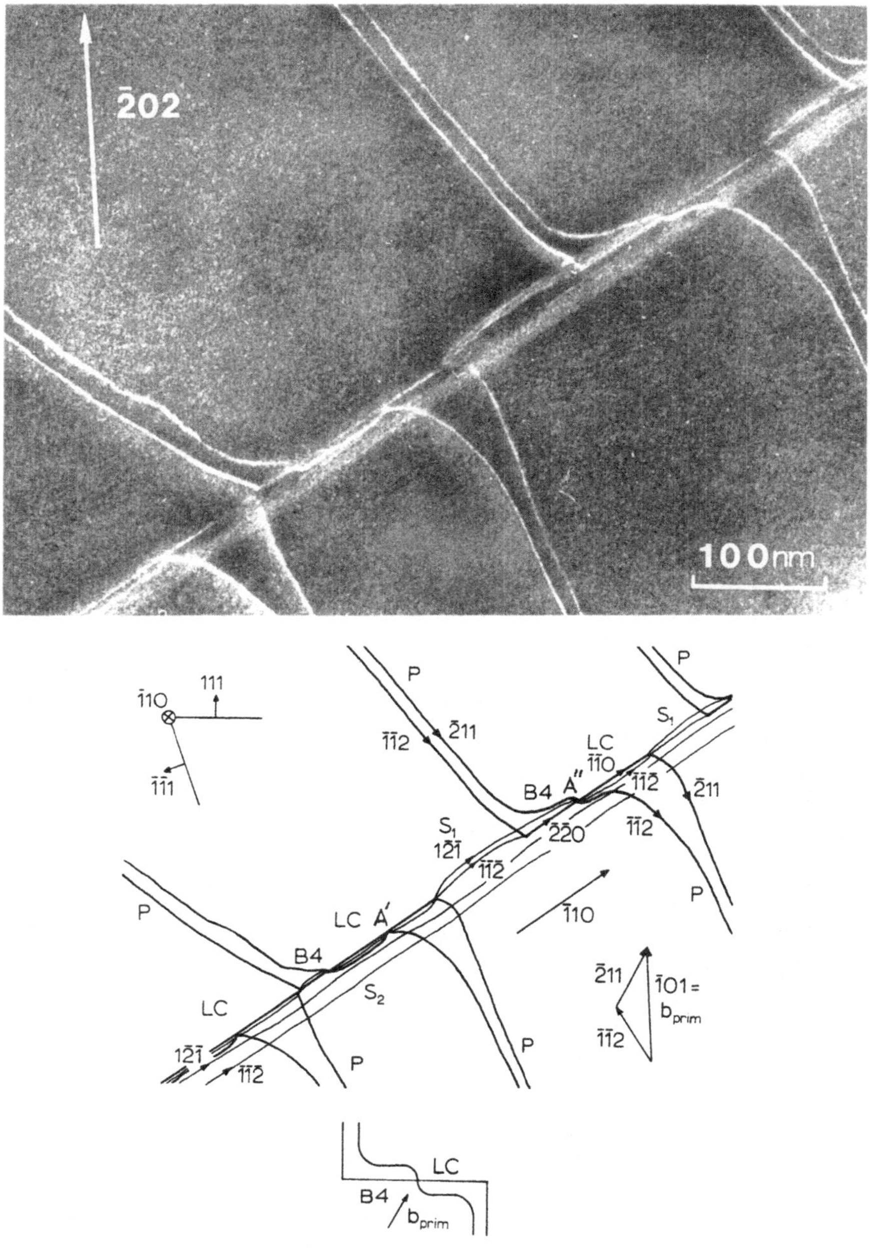

Fig. 10. Lomer–Cottrell barrier formation in Cu–15 at.%Al alloy. The micrograph is a weak-beam image. Burgers vector indices (multiplied by six) are given in the sketch.

Reprinted from *Acta Met.* **23**. H. P. Karnthaler and E. Wintner, Weak Beam TEM Studies of Composite Dislocations in F.C.C. Alloys, © 1975, Pergamon Press plc.

a small long-range attraction between parallel whole dislocations with Burgers vectors at 90°. The Hirth lock should readily un-zip into its constituent dislocations. In fact the Hirth locks studied by Karnthaler and Wintner[21] were observed at a uniformly spaced array of primaries where un-zipping is expected to be less favourable.

The Vienna group has also studied the glide of the undissociated Lomer dislocation on {001} in the fcc metals Ni, Cu and Ag,[23, 24] and the glide of bowed-out segments of dislocation on {001} in disordered Ni_3Fe alloy[25] and Cu–Al alloy.[26, 27] Karnthaler and Fischer[28] have found that the frequency of occurrence of Lomer–Cottrell locks does not correlate with the work hardening rate in stage II, for example in nickel, where the density of LC's is very low, the work hardening rate is high. The idealised model of piled-up loops behind parallelogram shaped LC barriers is not observed in practice. Furthermore, Karnthaler (private communication) suggests that the observed glide of Lomer dislocations on {001}, particularly in metals of high stacking fault energy such as Ni and Al, may explain the onset of stage III of the working hardening curve, since the primary, secondary and resultant dislocations have a common cross-slip plane.

Barriers in ordered alloys

There has been much interest recently in the behaviour of dislocations in ordered alloys of the $L1_2$ type, e.g. $Ni_3(Al,Ti)$, Ni_3Ga, etc. These alloys typically have a yield stress peak at high temperatures, and the softening above the peak is believed to be due to slip on {100} planes. Super dislocations in the ordered structure have twice the Burgers vector of the disordered structure, and they can dissociate in various ways. A super Lomer dislocation can dissociate to form a super Lomer–Cottrell lock as shown in Fig. 11(b). This is the analogue of the ordinary Lomer–Cottrell lock (Fig. 11(a)). Super intrinsic stacking fault (SISF) is involved, the sense of dissociation being opposite to that of Fig. 11(a) because Thompson's rule about the order of dissociation into partials for a given type of fault is reversed for a super dislocation. However, if the super Lomer dislocation dissociates first into two super edge partials on the {100} plane separated by antiphase domain boundary (APB), these can dissociate further as in Fig. 11(c). Complex stacking faults (CSF) on inclined {111} planes are thereby formed. Another possibility is that $B\alpha$ and $\alpha\beta$ (and $\alpha\beta$ and βA) combine to give Frank partials $B\beta$ (and αA). These are possible configurations for a 'complex' Lomer–Cottrell lock. Evidence for the super Lomer–Cottrell lock has been obtained by Sun and Hazzledine[29] for Ni_3Ga alloy deformed 3% at 700 °C. Figure 12 is a weak-beam image, believed to be such a super Lomer–Cottrell lock. The lock is constricted at either end,

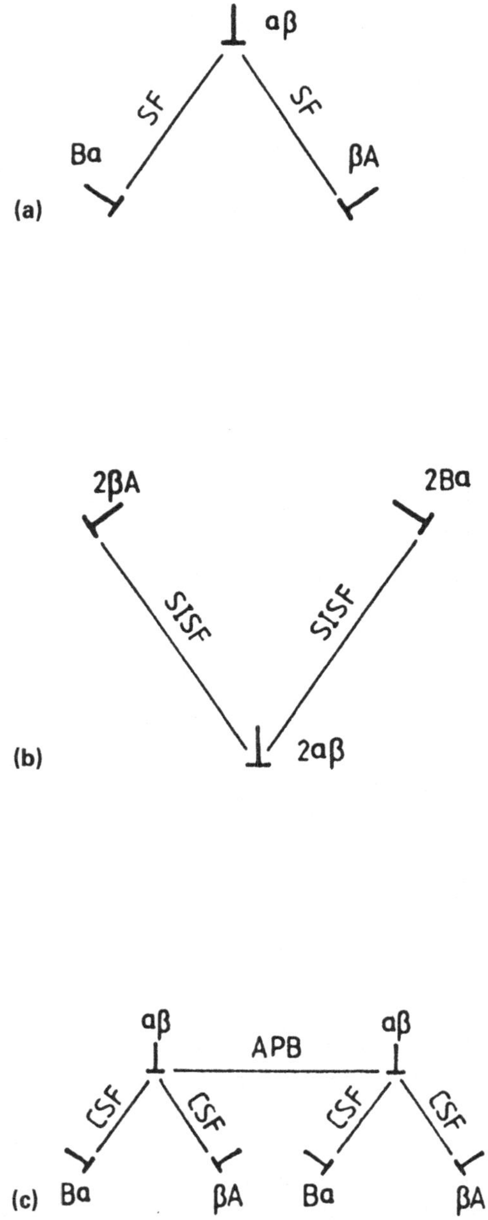

Fig. 11. (a), (b), (c). Possible Lomer–Cottrell dislocation dissociations. (a): Simple f.c.c. metal, Burgers vector BA, normal Lomer–Cottrell lock. (b): Super alloy, Burgers vector 2BA, super Lomer–Cottrell lock. (c): Super alloy, Burgers vector 2BA; 'complex' Lomer–Cottrell lock.
Sun and Hazzledine[29].

Fig. 12. Weak-beam image of a super Lomer–Cottrell lock in ordered Ni_3Ga super alloy deformed 3% at 700 °C.
Sun, unpublished work.

where the line direction changes to mixed character fairly close to screw orientation, as is expected for a super Lomer–Cottrell lock. It is seen that the mixed dislocations have only two super partials visible as expected, since further dissociation cannot occur for the mixed super partials.

A noteworthy point is that the super Lomer–Cottrell locks in the super alloys appear to form not by the dislocation interaction mechanism of Lomer[10] but by cross-slip of super screw dislocations from the close-packed plane to the cube plane, as originally proposed by Kear and Wilsdorf.[30] By this mechanism a bowed-out loop on the {001} plane is formed. The segments of such a loop at edge orientation are Lomer dislocations, and, as is visible in Fig. 12, the Lomer dislocations are elongated because of their sessile dissociation and the higher mobility of the screw parts.

Figure 13 (Sun, unpublished work) is a high resolution lattice image of a (110) foil of Ni_3Ga ordered single crystal deformed at 700 °C, which illustrates an interesting dissociation of a super Lomer edge dislocation by a climb mechanism different from the usual Cottrell glide mechanism. The edge dislocation along [110] is viewed end-on and is extended in the $(\bar{1}1\bar{1})$ plane as a super intrinsic stacking fault (SISF) and in the $(\bar{1}11)$ plane as a super extrinsic stacking fault (SESF). The Burgers vectors are indicated in the inset diagram, the dissociation being $2BA \rightarrow B\beta + \alpha A + \frac{2}{3}BA$. $B\beta$ and αA are Frank dislocations terminating the faults meeting at an obtuse angle. The complex has an asymmetric structure, the extrinsic fault being narrower than the intrinsic fault. This is probably due to the higher extrinsic stacking fault energy.

Finally we mention an interesting lattice image observation of Crimp[31] (Fig. 14) concerning the dissociation of a super screw dislocation in ordered Ni_3Al to form a Kear–Wilsdorf lock,[30] which acts as a barrier to the motion of screws, and which is thought to be responsible for the increase in yield stress with temperature in these alloys. The dislocation dissociates on the {100} plane forming APB, and each screw super partial may dissociate further into Shockley partials bounding a complex stacking fault on one of the two possible {111} planes containing the screw direction. In Fig. 14 the dislocation is viewed end-on. The two screw super partials dissociated on {111} are indicated with arrows and are connected by the invisible APB on the (010) plane whose trace is marked. The contrast is due to the edge components of the constituent Shockley partials, which are separated by only 3 or 4 atomic distances. The electron micrograph confirms the structure predicted for a Kear–Wilsdorf lock.

Conclusion

It is clear from what has been mentioned in this review that transmission

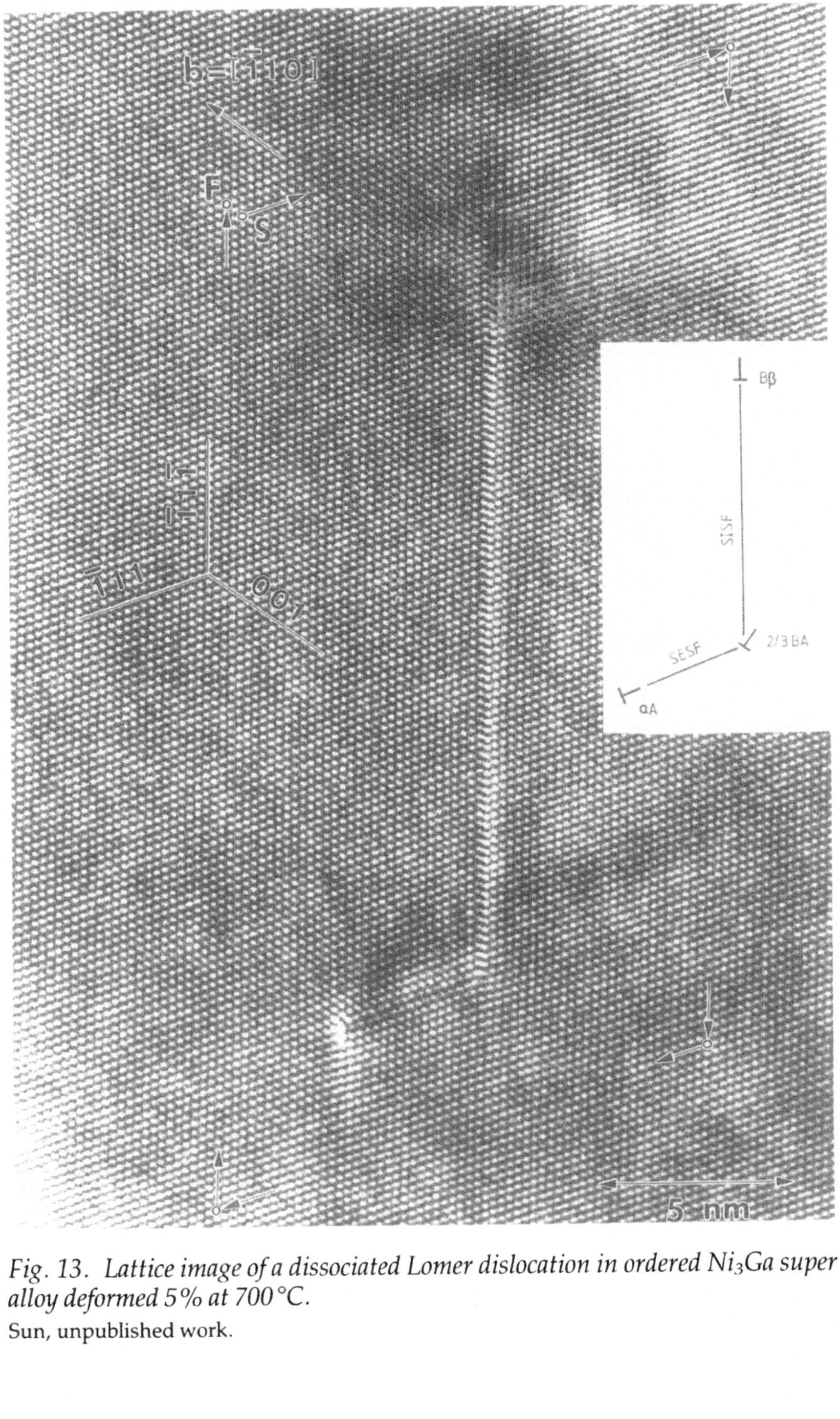

Fig. 13. Lattice image of a dissociated Lomer dislocation in ordered Ni_3Ga super alloy deformed 5% at 700 °C.
Sun, unpublished work.

Fig. 14. Lattice image of a super screw dislocation in ordered Ni_3Al which has dissociated to form a Kear–Wilsdorf lock (Crimp[31]).
Courtesy of the Philosophical Magazine.

electron microscopy has revealed a wealth of information about dislocation reactions and interactions. The Lomer–Cottrell lock and its variants is one of the most frequently observed resultant dislocations formed in the networks by interactions, although it does not appear in the form originally hypothesized, i.e. as extended barriers to primary pile-ups. These observations support models of work-hardening based on forest interaction, rather than those based on long-range stresses of pile-ups.

Acknowledgements

The authors wish to thank Dr P. M. Hazzledine, Professor H. P. Karnthaler and Mr Y.-Q. Sun for useful discussions, for kindly supplying illustrations and for allowing us to quote their unpublished work. We are also grateful to Professor D. J. H. Cockayne, Dr I. L. F. Ray and Dr M. A. Crimp for permission to reproduce micrographs.

References

1. A. Kelly, Applications of an X-ray Microbeam Technique, Ph.D Thesis, University of Cambridge, 1953.
2. P. B. Hirsch, A. Kelly and J. W. Menter, in *Proceedings of the Third International Conference on Electron Microscopy*, London, Royal Microscopical Society, pp. 231–235, 1954.
3. P. B. Hirsch, A. Kelly and J. W. Menter, *Proc. Phys. Soc.* **B68**, 1132–1145, 1955.
4. P. B. Hirsch, R. W. Horne and M. J. Whelan, *Phil. Mag.* **1**, 677–684, 1956.
5. I. L. F. Ray and D. J. H. Cockayne, *Proc. Roy. Soc.* **A325**, 543–554, 1971.
6. M. J. Whelan, *J. Electron Microscopy Technique* **3**, 109–129, 1986.
7. N. F. Mott, *Phil. Mag.* **43**, 1151–1177, 1952.
8. J. Friedel, *Phil. Mag.* **46**, 1169–1186, 1955.
9. A. Seeger, J. Diehl, S. Mader and H. Rebstock, *Phil. Mag.* **2**, 323–350, 1957.
10. W. M. Lomer, *Phil. Mag.* **42**, 1327–1331, 1951.
11. A. H. Cottrell, *Phil. Mag.* **43**, 645–647, 1952.
12. M. J. Whelan, P. B. Hirsch, R. W. Horne and W. Bollmann, *Proc. Roy. Soc.* **A240**. 524–538, 1957.
    W. Bollmann, *Phys. Rev.* **103**, 1588–1589, 1956. W. Bollmann, in *Proceedings of the European Conference on Electron Microscopy*, Stockholm, 1956, pp. 316–318, Stockholm, Almqvist and Wiksell, 1956.
13. M. J. Whelan, *Proc. Roy. Soc.* **A249**, 114–137, 1959.
14. A. Howie, in *Proceedings of the European Conference on Electron Microscopy*, Delft, 1960, Vol. I, 383–386; Delft, De Nederlandse Vereniging voor Electronenmicroscopie, 1961.
15. J. W. Steeds, *Proc. Roy. Soc.* **A292**, 343–373, 1966.
16. L. M. Brown, *Phil. Mag.* **10**, 441–466, 1964.

17. J. P. Hirth, *J. Appl. Phys.* **32**, 700–706, 1961.
18. T. Jossang, J. P. Hirth and C. S. Hartley, *J. Appl. Phys.* **36**, 2400–2406, 1965.
19. A. Korner, H. Schmid and F. Prinz, *Phys. Stat. Sol.* **51a**, 613–621, 1979.
20. D. J. H. Cockayne, I. L. F. Ray and M. J. Whelan, *Phil. Mag.* **20**, 1265–1270, 1969.
21. H. P. Karnthaler and E. Wintner, *Acta Met.* **23**, 1501–1509, 1975.
22. H. P. Karnthaler, P. M. Hazzledine and M. S. Spring, *Acta Met.* **20**, 459–468, 1972.
23. H. P. Karnthaler, *Phil. Mag.* **38**, 141–156, 1978.
24. A. Korner and H. P. Karnthaler, *Phil. Mag.* **44**, 275–284, 1981.
25. A. Korner and H. P. Karnthaler, *Phil. Mag.* **42**, 753–762, 1980.
26. H. P. Karnthaler and A. Korner, *Strength of Metals and Alloys*. Pergamon Press (ed. P. Haasen *et al.*), pp. 35–40, 1979.
27. A. Korner and H. P. Karnthaler, *Phys. Stat. Sol.* **75**, 525–535, 1983.
28. H. P. Karnthaler and L. Fischer, *Zeitschrift Metallk.* **66**, 631–638, 1975.
29. Y-Q. Sun and P. M. Hazzledine, in *Materials Research Society Symposium Proceedings* Vol. 133, on 'High temperature ordered intermetallic alloys', (ed. C. C. Koch, C. T. Liu, N. S. Stoloff and A. I. Taub), pp. 197–202. Pittsburgh, Pennsylvania, Materials Research Society, 1989.
30. B. H. Kear and H. G. F. Wilsdorf. *Trans. Met. Soc. AIME*, **224**, 382–386, 1962.
31. M. A. Crimp, *Phil. Mag. Lett.* **60**, 45–50, 1989.

5 The Nature of Grain Boundaries in Ordered Alloys

R. W. Cahn

Research Associate, Department of Materials Science and Metallurgy, University of Cambridge

Respectfully dedicated to Sir Alan Cottrell on the occasion of his 70th birthday

Abstract

Evidence concerning the structure of large-angle grain boundaries in ordered alloys such as Cu_3Au or Ni_3Al comes from simulation, high-resolution electron microscopy, mechanical property measurements and experiments on recrystallization. Taken together, the available information suggests that under certain circumstances the vicinity of such grain boundaries is disordered while the grain interiors remain ordered. This paper reviews the relevant evidence and why this phenomenon, if it indeed exists, might be of practical importance.

Introduction

During the past few years there has been intensive research activity, especially in the U.S., directed towards the possible use as load-bearing materials, at high temperatures, of refractory intermetallic compounds in polycrystalline form. Of such compounds — Ni_3Al, Fe_3Al, Ti_3Al, NiAl, TiAl, $TiAl_3$, Ni_4Mo are examples — many are ordered up to their melting-points. The main problem with many of these alloys is their characteristic brittleness at and near room temperature, which poses a grave obstacle to their application in jet engines and other critical uses. The brittleness of many of these alloys is attributable to the presence of grain boundaries; some of the alloys have good ductility in single crystal form. Other 'model' systems with a lower ordering energy disorder before they melt: the most studied such model system is Cu_3Au, others include FeCo and $(Co,Fe)_3V$, and such alloys are *not* brittle in polycrystalline form even when ordered.

Severely brittle alloys, for instance NiAl with the B2 structure, appear to lack ductility in polycrystalline form because of an insufficient number of slip systems, so that von Mises' celebrated criterion (for the ability of enclosed grains in a polycrystal to accept arbitrary deformation) predicts brittleness. Rachinger and Cottrell[1] were the first to analyse the factors which determine the favoured slip system: specifically, they analysed the effect of changes in the ordering energy on the relative energy of dislocations with different Burgers vectors. This pioneering study has stimulated a whole series of follow-up studies on B2 phases[2–8] which built on Rachinger and Cottrell's ideas, though they still leave unresolved the question why some alloys of this type change slip system as the temperature changes. It is clear that Rachinger and Cottrell's study was one of the motivating publications that nucleated the current avalanche of research on alloy phases such as NiAl.

One important issue which has been reopened by recent research is precisely the question, generally supposed to have been settled long ago, of the minimum number of 'independent' (i.e. non-redundant) slip systems required to permit a crystal grain, enclosed within a polycrystalline assembly, to adopt any arbitrary imposed homogeneous strain. Ever since a renowned study by Taylor,[9] based on theoretical work by von Mises over 60 years ago,[10] this number has been taken to be *five*. (A recent paper[11] in which the von Mises criterion is applies to deformation twinning, in possible combination with slip, shows the continued faith placed in this principle.) Fleischer[12] has reexamined this issue, by measuring the average number of participating slip systems in the grains of deformed polycrystalline α-brass. He found that 40% of the grains utilised 4 or fewer slip systems. He suggested that the von Mises criterion of a minimum of 5 non-redundant slip systems is not obeyed because the underlying assumption of uniform deformation, as he himself demonstrates, does not apply. Many metallographic and X-ray studies in recent years have established the notable heterogeneity of deformation in the grains of polycrystals (see an overview of the evidence by Cahn[13]), and it has thus become clear that a large number of non-redundant slip systems is not a necessary condition for polycrystalline deformability ... neither is their availability by any means a guarantee of such deformability! Just one recent example of the incidence of plastic deformability in a system with insufficient slip systems to satisfy the von Mises criterion is a B2 phase, Ni–30Al–20Fe, which slips only along $\langle 100 \rangle$ (3 independent slip systems) but has ≈2% tensile ductility.[14]

Recent research has made it very clear that not only the slip characteristics of the constituent grains, but also the nature of the grain boundaries, influence the brittleness or ductility of alloys such as Ni_3Al and NiAl. Thus, whereas [B2] NiAl is brittle, [B2] FeCo can have extensive ductility, and this requires explanation. The aim of this short paper is to give an overview of recent research which has a bearing on this category of question.

Constitutional Disordering

The intensive researches during the past few years on polycrystalline Ni_3Al and its highly alloyed derivatives derive directly from the discovery, in 1979, that small doping additions of boron turn this brittle material into a ductile one, provided that the nickel content is slightly above 75 at.%. The efficacy of boron in turn derives from its strong segregation to the grain boundaries. This much is beyond any reasonable doubt; the evidence has been presented in many papers and is summarised in a recent review of the physical metallurgy of Ni_3Al.[15] The next stage of the argument, still subject to some conflicting experimental findings, is that the segregated boron draws in nickel from the grains (provided the initial nickel content exceeds 75 at.%), so that the nickel content of the grain boundaries is enhanced. The last stage of the argument is that this nickel enrichment, in turn, locally weakens the ordering energy and this can bring about a narrow zone adjacent to grain boundaries where the superlattice disappears.

The evidence concerning the linked B and Ni segregation comes mainly from a very detailed study[16–19] by Baker, Schulson and coworkers of Dartmouth College, using both X-ray spectra and CBED studies in TEM; similar findings were reported by Krzanowski, using X-ray microanalysis.[20] A careful critique of the contrary evidence by George *et al.*[21] will be found in one of the Dartmouth College papers.[18] In the latest of the Dartmouth College papers, Ni contents in the grain boundary zone of a $Ni_{76}Al_{24}$ alloys are reported as exceeding 95 at.%! Baker, Schulson and Michael[16] in fact predicted that such grain boundaries of modified composition should locally lose atomic order.

Direct evidence concerning compositional disordering comes originally from a HREM study by Mackenzie and Sass[22] on grain boundaries in $Ni_{76}Al_{24}$ doped with 300 ppm of B and equilibrated at 1000 °C. A change in the lattice fringe contrast in a narrow zone, 1·5–2·5 nm wide, was interpreted as demonstrating the absence of atomic order there. This observation confirmed a number of earlier conjectures and less direct experimental indications, which are reviewed by Mackenzie and Sass. At least one subsequent attempt to confirm these HREM findings was reported to have been unsuccessful and Mills[23] has challenged the interpretation of the HREM contrast in terms of atomic disorder, claiming that this type of contrast is ultrasensitive to grain orientation and can thus give a misleading impression. However, Schulson (refs. 17 & 18 and an oral presentation at the Symposium on Aluminides, Indianapolis, 4 October, 1989) has not only directly confirmed the nickel enrichment but also metallographically demonstrated the presence of an epitactic, disordered phase up to 20 nm wide at many grain boundaries. Schulson *et al.*[18] established that the nickel enhancement was pronounced and universal on all grain boundaries examined *only* if the nickel content exceeded 75 at.% and boron was present. Schulson has proposed the term 'constitutionally disordered zone round grain boundaries' to describe his observations.

It is to be presumed that the presence of constitutional disordering at grain boundaries is contingent not only on bulk composition (and doping) but also on the temperate of equilibration, and it is intuitively likely that the phenomenon would be favoured by a high equilibration temperature, fairly close to the (virtual) disordering temperature, which ≈ 1450 °C for stoichiometric Ni_3Al[24], ≈1390 °C (=the melting temperature) for $Ni_{77}Al_{23}$ and below the melting temperature for higher nickel contents. Up to now, only an equilibration temperature of 1000 °C has been used. It would be particularly interesting to establish whether a less strongly ordered $L1_2$ phase with a critical temperature for disordering, T_c, well below the melting temperature shows the presence of constitutional disordering near grain boundaries when it is equilibrated at a temperature not far below T_c, and experiments to establish this are currently under way (J. M. Howe and R. W. Cahn).

As to the enrichment of grain boundaries in the majority constituent, Ni in the case of Ni_3Al, there is no corresponding evidence relating to other $L1_2$ phases. There is a body of experimental evidence concerning such enrichment at *free surfaces* in Cu_3Au and there is a clear theoretical indication of enrichment (and concomitant reduction of the order parameter) in a zone a few atoms wide adjacent to *antiphase domain boundaries;*[25] this is reinforced by experimental evidence that films of disordered phase are preferentially nucleated at APDBs (in Ni_3Al + Fe) on heating the alloy.[26, 27] The evidence concerning surfaces and APDBs as it stood three years ago is fully set out in Ref. 26. One can attempt a prediction of the likely level of equilibrium segregation to grain boundaries of an A-rich A/B alloy in terms of well established general relationships (see Hondros and Seah[28]): the degree of enrichment is inversely related to the solubility, and since the solubility of Ni in Al is very much smaller than that of Al in Ni, one would predict Ni enrichment at grain boundaries. There is no general theory to indicate how the presence of a superlattice should modify equilibrium segregation. Incidentally, Hondros and Seah demonstrate the close relation between equilibrium segregation to grain-boundaries and to surfaces.

Chen and his collaborators at Los Alamos have made extensive first principles calculations on the behaviour of boron at grain boundaries and free surfaces in Ni_3Al, and have been able to confirm[29] the experimental finding that boron segregates more strongly to grain boundaries than to free surfaces, that boron and nickel cosegregate to grain boundaries, and that this effect is much stronger for Ni-rich boundaries.

Farkas and Jang[30] performed a CVM computation of the behaviour of a $\Sigma = 5$ grain boundary for a two-dimensional lattice gas model, and concluded that a disordered layer forms in the grain boundary region at temperatures significantly below the bulk T_c.

There is one other, indirect, source of evidence concerning a possible constitutionally disordered zone adjacent to grain boundaries in ordered alloys. This is the formation of a network of antiphase domains in grains formed by primary recrystallisation after plastic deformation. This is relevant because APDs can only form during cooling of the disordered alloy (for evidence see Ref. 26), and thus the formation of APDs during recrystallisation implies that the advancing grain boundaries carry with them a disordered zone. Recent research on Cu_3Au[31,32] has established that APD networks are indeed formed on recrystallising at a temperature 30 °C below T_c; however, Ni_3Al grains formed by recrystallisation contain hardly any APDs. So far as this somewhat disputable line of evidence goes, Ni_3Al should not show constitutional disordering (at most grain boundaries), whereas Cu_3Au should.

Implications of Constitutional Disordering for Mechanical Properties

The observation of constitutional disordering in B-doped Ni_3Al by Mackenzie and Sass led Baker and Schulson[33] to a most intriguing hypothesis. They start from the fact that for a number of reversibly ordered polycrystalline alloys (Ni_3Mn, Ni_3Fe, Cu_3Au and FeCo) the Hall–Petch slope is much steeper for the alloys in their ordered form than in the disordered form. The ratio of slopes in the two conditions range from 1.35 to 3.5. As they point out, this slope is determined by the 'contribution of the grain boundaries to the yield strength', though they concede that changes in slip character within the grains will also have an effect. They go on to postulate that a grain boundary in (disordered) fcc or bcc alloy will transmit slip more easily than a grain boundary in a superlattice such as $L1_2$ or B2 (and also that the former type will dissipate stress concentrations more effectively through the motion of grain-boundary dislocations): it follows that an alloy which has ordered grains but a disordered zone around grain boundaries should have a less steep Hall–Petch slope

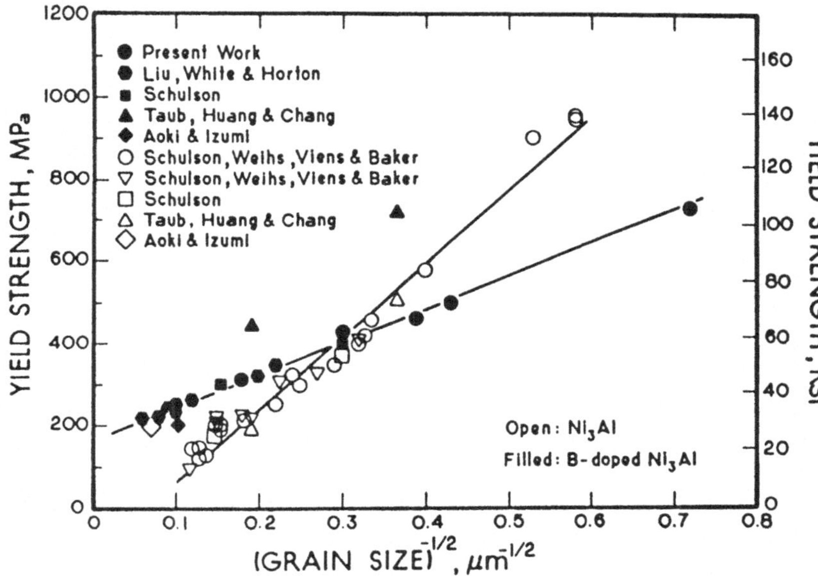

Fig. 1. Comparison of Hall–Petch relationships for Ni_3Al with and without boron dopant (After Ref. 34).

and behave in a more ductile fashion. Figure 1 (from Khadkikar *et al.*[34]) claims to show that B-doped Ni_3Al has a lower Hall–Petch slope than the undoped alloy; if indeed the doped alloy has constitutionally disordered grain boundaries, then this would be consistent with the notion that it is primarily the state of the grain boundary, not that of grain interiors, which determines the Hall–Petch slope and the mechanical properties associated therewith.* In an even more recent paper, Baker *et al.* (Ref. 19) examine the behaviour of Ni_3Al and Ni_3Si in terms of their hypothesis and find general agreement.

In a recent study, Cheng *et al.*[35] irradiated undoped $Ni_{76}Al_{24}$ with high-energy protons and found that this partially disordered the material. The fractographs of the partially disordered zones indicated local ductility, in contrast with the total grain-boundary brittleness of the

---

* The particular plot for undoped Ni_3Al shown in Fig. 1 has given rise to much argument because the intercept for infinite grain diameter comes at a negative yield stress. In his overview, Stoloff[15] analyses the full range of published Hall–Petch measurements for Ni_3Al and concludes that B has little effect on the Hall–Petch slope for 25 at.% Al but reduces the slope by about a quarter for an alloy with 24 at.% Al. This is fully consistent with the Baker/Schulson hypothesis.

regions which had remained fully ordered. This again confirms the ductilising effect of local disorder in this alloy.

Others have sought to interpret the change in ductility with doping of Ni_3Al in terms of grain-boundary cohesion. In particular, Chen and coworkers[29] have proposed a carefully constructed theory on these lines, based on first-principles calculation of boundary cohesion, for details of which their original papers should be consulted. Another approach which may have relevance to the problem has recently been proposed by Cottrell,[36] though his theory was originally applied to pure metals. According to this model, grain boundary weakness or strength can be related to the ratio of shear modulus to bulk modulus in the crystals. Cottrell concluded that the boundary is weaker than the crystal in a wide range of metals, favouring intercrystalline fracture in preference to cleavage. The grain boundary becomes relatively more resistant to fracture as the ratio of shear modulus (for the slip direction and plane) to bulk modulus, (μ/K), *decreases*. He demonstrates this for a range of pure metals.

The elastic (crystal) moduli for three $L1_2$ superlattices (Ni_3Al, Ni_3Fe and Cu_3Au) at room temperature are cited by Yoo.[37] However, data for the change of moduli with degree of order are scarce, and the only ones known to me were determined for Cu_3Au[38] and CuZn[39,40] 50 years ago. Generally, alloys become stiffer when they become ordered.[41] It is not possible to measure elastic properties for disordered Ni_3Al since the only way to disorder this compound completely is by extreme cold work (filing); quenching will not serve. It is possible to make calculations along the lines of Cottrell's theory for Cu_3Au: Siegel[38] measured the elastic constants c_{11}, c_{12} and c_{44} as a function of temperature, using single crystals of this compound. At the critical temperature, T_c (=387·5 °C according to Siegel), there is a discontinuous change of the elastic constants (Fig. 2): on cooling through T_c, the order parameter increases discontinuously from 0 to ≈0·8, and (c_{11}—c_{12}) increases discontinuously by 10% while c_{44} increases by 2%. From this, values of μ/K at T_c can be calculated, and it turns out that μ/K for the ordered alloy is $0{\cdot}18_5$ while for the disordered alloy the ratio is $0{\cdot}16_9$. The reduction is by 8·6%. In fact, the weighted average of μ/K for elementary Cu and Au is $0{\cdot}19_6$; the effect of alloying and ordering here is comparatively slight. Values of μ/K around 0·18 are characteristic of highly ductile metals,[36] and indeed Cu_3Au in both ordered and disordered forms is highly deformable.

The elastic constants cited by Yoo[37] for Ni_3Al permit a value of μ/K of $0{\cdot}28_3$ to be calculated for the ordered alloy at room temperature; this is still not a very high value when compared with values for pure metals such as platinum (0.237) which is certainly ductile, or nickel (0.343) which is ductile when pure but is very sensitive to grain-boundary embrittlement

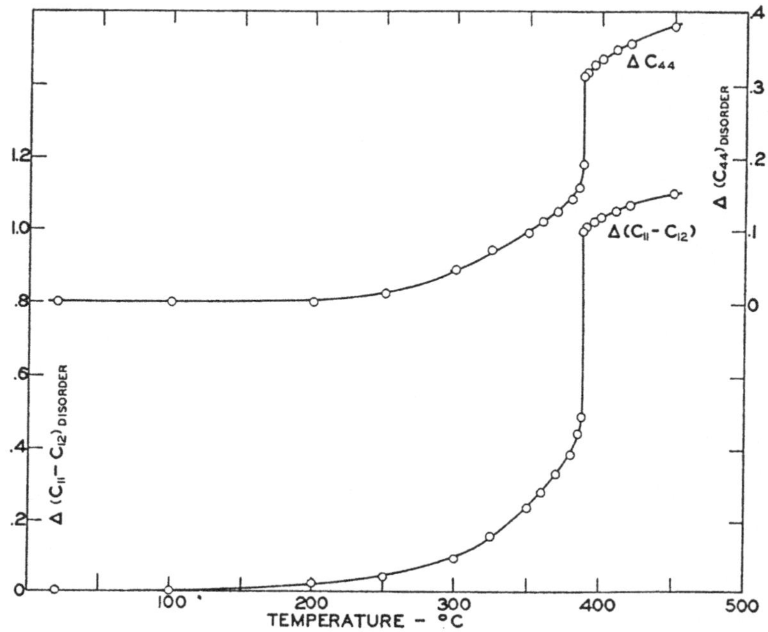

Fig. 2. The decrease *(relative to room-temperature values) of the elastic constants of Cu_3Au plotted against the temperature of measurement. The discontinuous change at* T_c *is clearly apparent (After Ref. 38).*

by contaminants. It would be very interesting to see by how much Young's modulus changes when a derivative of Ni_3Al is disordered . . . it will not be feasible to measure individual elastic constants! This will have a bearing on the interpretation of Baker and Schulson's model for ductilisation of intermetallics by the creation of a constitutionally disordered grain-boundary zone.

Acknowledgements

I am grateful to Prof. I. Baker, Dr S. P. Chen, Sir Alan Cottrell and Dr. M. J. Mills for the provision of copies of their latest papers in advance of publication.

References

1. W. A. Rachinger and A. H. Cottrell, *Acta Metall.* **4**, 109–113, 1956.
2. A. Ball and R. E. Smallman, *Acta Metall.* **14**, 1517–1526, 1966.

3. E. P. Lautenschlager, T. Hughes and J. O. Brittain, *Acta Metall.* **15**, 1347–1357, 1967.
4. D. I. Potter, *Mat. Sci. & Eng.* **5**, 201–209, 1969–70.
5. Y. Umakoshi and M. Yamaguchi, *Phil. Mag. A.* **41**, 573–588, 1980.
6. M. G. Mendiratta and C. C. Law, *J. Mater. Sci.* **22**, 607–611, 1987.
7. P. R. Munroe and I. Baker, *Scripta Metall.* **23**, 495–499, 1989.
8. I. Baker, Paper presented to the Electron Microscopy Society of America, August 1989, to be published in the proceedings of the 1989 meeting.
9. G. I. Taylor, *J. Inst. Metals* 1934, 307 (1934).
10. R. von Mises, *Z. angew. Math. Mech.* **8**, 161, 1928.
11. E. Goo and K. T. Park, *Scripta Metall.* 1053–1056, 1989, 1989.
12. R. L. Fleischer, *Acta Metall.* **35**, 2129–2136, 1987.
13. R. W. Cahn, *Phil. Trans. R. Soc. Lond. A.* 155–176 (1978), 1978.
14. S. Guha, P. Munroe and I. Baker, *Scripta Metall.* **23**, 897–900, 1989.
15. N. S. Stoloff, *Int. Mater. Reviews.* **34**, 153–183, 1989.
16. I. Baker, E. M. Schulson and J. R. Michael, *Phil. Mag.* B. **57**, 379, 1988.
17. I. Baker and E. M. Schulson, *Scripta Metall.* **23**, 1883–1886, 1989.
18. I. Baker, E. M. Schulson, J. R. Michael and S. J. Pennycook, *Phil. Mag.* B. In press, 1990.
19. I. Baker, E. M. Schulson, J. R. Michael and R. A. Padgett, *J. de Physique*, in press.
20. J. E. Krzanowski, *Scripta Metall.* **23**, 1219–1224, 1989.
21. E. P. George, C. T. Liu and R. A. Padgett, *Scripta Metall.* **23**, 979–982, 1989.
22. R. A. D. Mackenzie and S. L. Sass, *Scripta Metall.* **22**, 1807–1812, 1988.
23. M. J. Mills, *Scripta Metall.* **23**, 2061–2066, 1989.
24. R. W. Cahn, P. A. Siemers, J. E. Geiger and P. Bardhan, *Acta Metall.* **35**, 1737–2751, 1987.
25. J. M. Sanchez, S. Eng, Y. P. Wu and J. K. Tien, *Mat. Res. Soc. Symp. Proc.* **81**, 57–64, 1987.
26. R. W. Cahn, P. A. Siemers and E. L. Hall, *Acta Metall.* **35**, 2753–2764, 1987.
27. R. W. Cahn and W. L. Johnson, *J. Mater. Res.* **1**, 724–732, 1986.
28. E. D. Hondros and M. P. Sheah, in *Physical Metallurgy*, 3rd edition (ed. R. W. Cahn and P. Haasen), 878–880. North-Holland Physics Publishing, Amsterdam, 1983.
29. S. P. Chen, A. F. Voter, R. C. Albers, A. M. Boring and P. J. Hay, *Scripta Metall.* **23**, 217–222, 1989; also a fuller treatment by S. P. Chen, D. J. Srolovitz and A. F. Voter, *J. Mater. Research* **4**, 62–77, 1989.
30. D. Farkas and H. Jang, *Mat. Res. Soc. Symp. Proc.* **81**, 65–72, 1987.
31. K. H. Westmacott and R. W. Cahn, to be submitted to *J. Mater. Sci.*
32. R. W. Cahn, in *High Temperature Aluminides and Intermetallics* (ed. S. H. Whang, C. T. Liu and D. Pope). Warrendale, PA, The Minerals, Metals and Materials Society of AIME, in press, 1990.
33. I. Baker and E. M. Schulson, *Scripta Metall.* **23**, 345–348, 1989.
34. P. S. Khadkikar, K. Vedula and B. S. Shabel, *Metall. Trans.* **18A**, 425–428, 1987.
35. J. Cheng, C.-S. Lee, C. N. J. Wagner and A. J. Ardell, *Mat. Res. Soc. Symp. Proc.* **133**, 499–504, 1989.

36. A. H. Cottrell, *Mat. Sci. Tech.* **5**, 1165–1167, 1989.
37. M. H. Yoo, *Mat. Res. Soc. Symp. Proc.* **81**, 207–212, 1987.
38. S. Siegel, *Phys. Rev.* **57**, 537–545, 1940.
39. J. S. Rinehart, *Phys. Rev.* **58**, 365–371, 1940.
40. W. A. Good, *Phys. Rev.* **60**, 605–609, 1941.
41. L. Guillet and R. le Roux, in *Intermetallic Compounds* (ed. J. H. Westbrook) pp. 453–463. New York, Wiley, 1967.

6 Radiation Damage and Point Defects

R. Bullough

Chief Scientist, United Kingdom Atomic Energy Authority

Abstract

A brief review of the many seminal contributions of Sir Alan Cottrell to our understanding of Radiation Damage and Point Defects is presented. These contributions refer to the response to irradiation of both fissile and non-fissile materials. An attempt is made to identify the relevance of his contributions, mostly made in the mid-fifties, to our present understanding of the phenomena.

1. Introduction

It is both an honour and a great pleasure to review briefly the many seminal contributions of Sir Alan Cottrell to our understanding of radiation damage and associated point defect properties in crystalline materials. My pleasure in doing this stems from my wish to acknowledge all the help and inspiration he has given to me since the day I met him, when together with Bruce Bilby (I was Bilby's first research student at Sheffield), I visited Birmingham to discuss my progress on one of Bilby's difficult dislocation theory problems.* He was so kind, tolerant and helpful, considering my relative ignorance — a state of affairs that I might say has continued to the present day!

Although his specific contributions to the science of radiation damage in both fissile and non-fissile materials were largely made during his period (1955–58) as Deputy Head of the Metallurgy Division at Harwell, it should perhaps be emphasized that his earlier work at Birmingham on the theory of yielding and strain ageing[1] has also been of profound relevance to much of our current understanding of the detailed microstructural changes that can take place when such materials are subject to high energy nuclear radiation. Cottrell's influence on nuclear issues has,

* As I recall we had derived the integral equation for the Peierls representation of a tilt boundary or wall of edge dislocations. This equation was non-linear, singular and of a non-standard form — needless to say we never did solve it.

of course, continued since his departure from Harwell; he has, for example, served two terms as a part-time Member of the UKAEA (1962–65 and 1983–86) and it was largely through his concern that the extensive and important investigations of the factors that influence the integrity of reactor pressure vessels were undertaken by the UKAEA and which culminated in the 'Marshall Report' in 1982,[2] followed by its recent sequel under Professor Sir Peter Hirsch.[3]

I should confess at the outset that the identification of his many specific contributions to radiation damage has been greatly facilitated by reading the text of the 46th Thomas Hawksley Lecture[4] which Cottrell prepared in 1959 entitled 'Effect of Nuclear Radiation on Engineering Materials'. These contributions take the form of enlightening experimental and theoretical discussions of radiation hardening, radiation embrittlement, radiation growth, creep and swelling of fissile and non-fissile materials. His various contributions to our understanding of these phenomena are described in Section 2. We also give a brief indication, in each case, as to whether his views have stood the test of time including, of course, the impact of the development of much improved observational techniques.

2. Specific Contributions

In this Section we present a brief review of the contributions made by Cottrell and his collaborators to our understanding of the various important radiation damage phenomena. Most of this research was performed during the period Cottrell spent at Harwell.

2.1 *Radiation hardening*

Following the early discovery that neutron irradiation does produce hardening (an increase in the yield stress) it was clear to Churchman, Mogford and Cottrell[5] in 1957 that such hardening could only be due to either

(i) the irradiation produced point-defects (interstitials or vacancies) pinning the existing dislocations; so called 'source hardening' in which the propagation of yield from grain to grain is inhibited by the inability of existing dislocation network segments to act as Frank–Read sources, or

(ii) the irradiation produced point-defects forming small aggregates or clusters dispersed throughout the lattice; so-called 'dispersion hardening' in which the propagation of yield is inhibited by friction from these clusters on the motion of dislocations.

These authors measured the hardening in a structural steel as a function of neutron dose and grain size and used the Hall–Petch relation for the dependence of the yield stress σ_y on grain size d

$$\sigma_y = \sigma_i + \kappa_y d^{-\frac{1}{2}}$$

where σ_i is the friction stress for dislocation glide across a grain and κ_y is a parameter proportional to the stress required to operate dislocation sources in adjacent grains, to deduce that only the friction stress σ_i increases with radiation dose. The parameter κ_y does not change with radiation dose and thus dislocation pinning is not a significant source of hardening; the contemporary results of Hull and Mogford[6] are shown in Fig. 1 where the invariance of slope with $d^{-\frac{1}{2}}$ is clearly apparent.

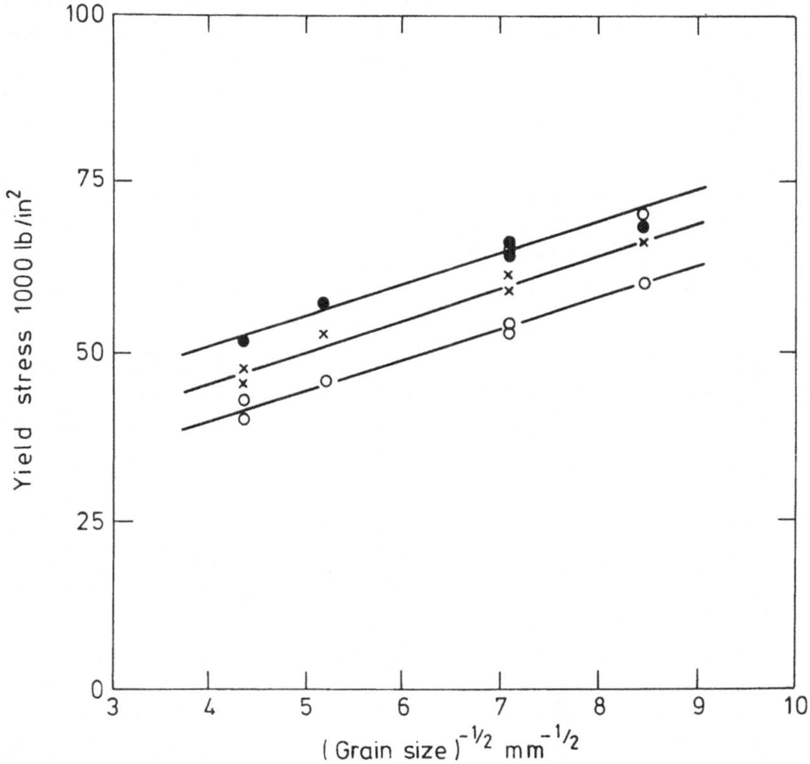

Fig. 1. The variation at −78 °C of yield stress with grain size of En2 steel before and after neutron irradiation. • *Irradiated* 3×10^{18} *neutrons* cm^{-2}; × *Irradiated* 1×10^{18} *neutrons* cm^{-2}; ○ *Unirradiated.*

(Reprinted from Hull and Mogford,[6] 1958 and referred to by Cottrell.[4])

The use of the Hall–Petch relation to deduce the mechanism of radiation hardening was undoubtedly an important milestone and led the authors to basically the correct conclusion that radiation hardening is indeed due to an increase in barriers to the glide of mobile dislocations and not to an increase in the difficulty of generating mobile dislocations. However, needless to say, in the intervening 32 years, the subject has become very complex and it will have to suffice to indicate merely the richness of the present scene and refer the interested reader to one of the modern reviews such as that by Little.[7]

Thus:

(i) there are complex impurity effects. The magnitude of radiation hardening increases with increasing interstitial impurity concentration. Irradiation leads to a depletion of such solute atoms from solution; such a loss of carbon and nitrogen from solution in iron is clear from the drop in capacity to strain age measured by a corresponding drop in the Snoek Peak.[8] The solute atoms increase the density of clusters and correspondingly reduce the solute capacity to pin the dislocations. At high doses when cluster nucleation has effectively ended the hardening increase is largely due to the continued segregation of the solute at the (now larger) clusters. Such irradiation enhanced segregation of solute is now known to be of enormous importance and this example serves to illustrate the importance of microchemical as well as microstructural evolution in any understanding of radiation damage processes.

(ii) the intrinsic clusters that lead to radiation hardening are now known to be small dislocation loops which are very often associated with segregated solute and the solute can both enhance the density of the clusters and increase their effectiveness as barriers to dislocation glide.[9] Copper has long been known[10] as an impurity that produces a very marked increase in the radiation hardening of both pure iron and low alloy steels. It is now considered[11] that the specific hardening effect of copper arises from the homogeneous nucleation and growth of very small copper precipitates during irradiation. The migration of the substitutional copper is irradiation enhanced by the mobile vacancies which both provide the means for the copper to migrate and provide the space to accommodate the copper precipitates. A recent high resolution TEM is shown in Fig. 2 together with definite EDX evidence for the presence of these very small copper precipitates. In addition to this hardening due to the presence of such coherent precipitates such metals

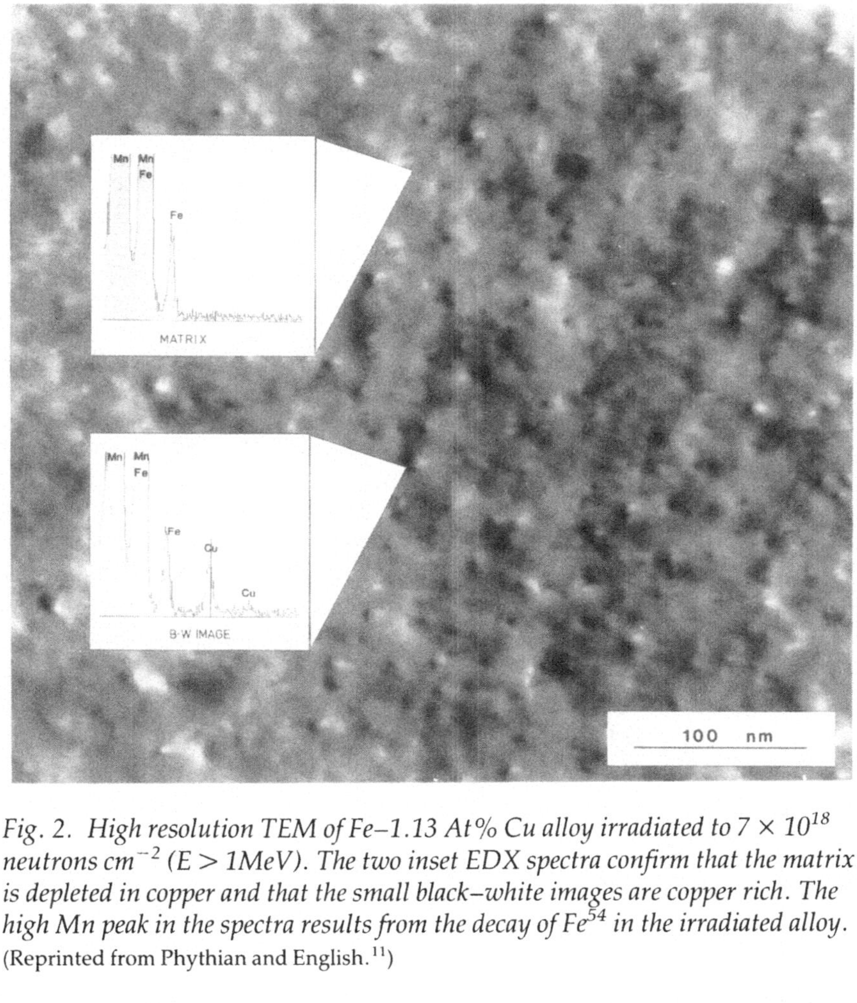

Fig. 2. High resolution TEM of Fe–1.13 At% Cu alloy irradiated to 7×10^{18} neutrons cm^{-2} ($E > 1MeV$). The two inset EDX spectra confirm that the matrix is depleted in copper and that the small black–white images are copper rich. The high Mn peak in the spectra results from the decay of Fe^{54} in the irradiated alloy. (Reprinted from Phythian and English.[11])

will also have a smaller component of hardening due to the intrinsic clusters (with or without associated solute!).

(iii) the realization that the hardening clusters can be dislocation loops and irradiation produced coherent precipitates has led to the notion of dispersed barrier hardening as the meaningful description of radiation hardening.[7] Thus, the penetrability of the obstacles has been introduced to the expression for the force to cause a dislocation to bow round an impenetrable obstacle. If N is the density of such obstacles and ℓ is their average size, then the increase in yield stress $\Delta\sigma_y$ due to their presence has the form

$$\Delta\sigma_y = \alpha\mu b(N\ell)^{-\frac{1}{2}} \qquad (1)$$

where μ is the shear modulus and b is the modulus of the Burgers vector of the dislocations; the quality α, which varies from zero to unity, represents the obstacle strength, and we note this increase is not very sensitive to temperature. It is now possible to see why certain solutes when segregated on the loops can so strongly affect the resulting hardening. However, the precise way that the glide dislocations can 'sweep' channels through dense colonies of dislocation loops is certainly not completely understood.

(iv) in situations when there is little effect of neutron dose rate it seems highly likely that the clusters that are responsible for hardening do not form by diffusion processes. Hence, it is probable that the vacancy loops formed by the athermal collapse of cascades are the essential cluster. However, the sensitivity to interstitial impurities is difficult to understand. It could be that they stabilize the vacancy loops and prevent their shrinkage. Alternatively, they could trap the mobile interstitials formed at the periphery of the cascades and thereby inhibit recombination with the vacancy loops. Finally, perhaps the trapped interstitial clusters are themselves effective hardeners. There is evidence that such solute segregation to a growing interstitial loop can stop its growth and cause the further nucleation of new interstitial loops to form stable colonies of such loops.[12] More clarification regarding the dose rate sensitivity would seem to be important. It perhaps should be emphasised that the hardening due to copper does appear to be sensitive to dose rate;[13] a fact that is consistent with the notion of irradiation enhanced diffusion of the copper.

2.2 *Radiation embrittlement*

The work of Churchman, Mogford and Cottrell[5] was again the first study to provide a mechanism for radiation embrittlement. Thus they posed the question: is the embrittlement caused by the hardening or has there been some mysterious loss of cohesive strength? Again, they exploited the Hall–Petch relation, but this time between the fracture stress σ_f and grain size

$$\sigma_f = \sigma_i + \kappa_f d^{-\frac{1}{2}}$$

where now the parameter σ_f measures the 'Stroh' stress; that is the stress needed to crack the material at the tip of a slip-band or pile-up of dislocations. Their results for the fracture energy absorbed using notched impact specimens of a mild steel with a particular grain size in the unirradiated and irradiated condition are shown in Fig. 3. Very considerable embrittlement, with a large increase in the temperature of brittle fracture arising from relatively low neutron dose is apparent in this steel. By measuring σ_f as a function of irradiation for the two widely differing grain sizes of the steel it was clear that the irradiation has negligible effect on the grain size dependence of σ_f and thus the authors concluded that the irradiation had not changed the cohesive strength of the metal (σ_f did not change) and therefore the shift in transition temperature in Fig. 3 is entirely due to the combined effect of change of yield stress with irradiation dose and temperature. At a given temperature the irradiation leads to an increase in yield stress due to the creation of obstacles for dislocation glide and the yield stress for both the irradiated and unirradiated state both fall with increasing temperature; on the other hand, the cohesive strength (cleavage strength) varies little with radiation or temperature and therefore

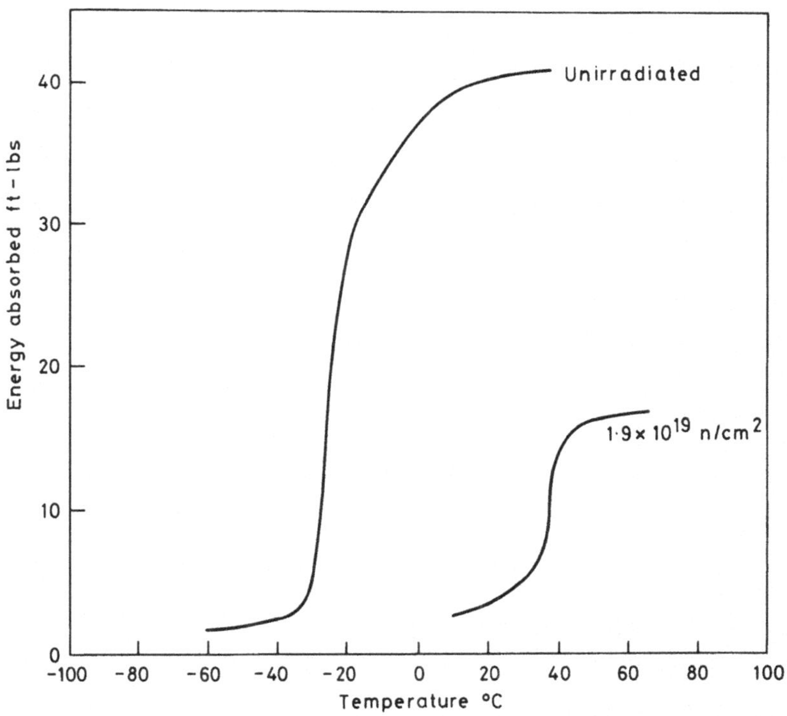

Fig. 3. The effect of neutron irradiation (1.9×10^{19} neutron cm^{-2}) on the DBTT transition temperature of En2 steel.

(Reprinted from Cottrell's 46th Thomas Hawksley Lecture,[4] 1959 from the original work of Churchman, Mogford and Cottrell,[5] 1957.)

brittle fracture will occur up to a higher temperature after irradiation — this is the origin of the shift in the so-called DBTT (Ductile Brittle Transition Temperature).

Whilst our understanding of radiation embrittlement, particularly in relation to the effects of solute elements such as copper etc., has improved in detail since this work by Cottrell and his colleagues, their conclusion that it has its fundamental origin in the hardening processes induced by irradiation has not changed. Figure 4 shows recent evidence of the dramatic effect of copper on the shift of the DBTT with irradiation dose. The only word of caution that is probably necessary in making this assertion is the likelihood that irradiation induced segregation[15–17] of 'dangerous' embrittling elements like sulphur or phosphorous to existing microcracks or grain boundaries could well cause significant changes in the cohesive strength of materials and perhaps radically change the notion that the cohesive strength does not change with irradiation dose.

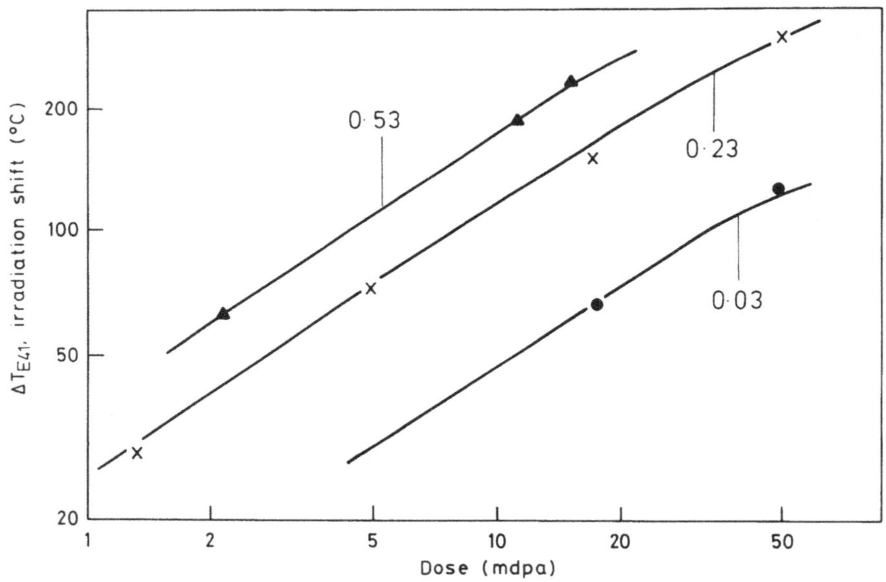

Fig. 4. The effect of neutron dose on the shift of the DBTT at 41 J energy for high nickel welds of A533B after irradiation at 250 °C. The numbers associated with the data points indicate the copper concentration in the welds in wt%.
(Reprinted from Williams *et al.*[14])

2.3 *Radiation growth and creep of materials*

Cottrell's specific contributions under this heading include theories of the radiation growth and creep[18] of uranium. It is also appropriate to acknowledge his important contribution[19] to our understanding of irradiation produced point-defects in graphite and the complex problem of the Wigner energy release in air-cooled graphite moderated reactors such as BEPO at Harwell. This latter success he recently referred to as 'This little adventure in solid-state engineering (which) passed off safely and successfully'.[20]

2.3.1 Radiation growth in uranium and non-fissile materials

α-uranium is fissile and has an orthorhombic structure. Under irradiation a single crystal undergoes a change of shape; it expands along the [010] (b-axis), contracts by an equal fraction along the [100] (a-axis) and remains unaltered along the [001]. At temperatures below about 400 °C there is little associated change of volume (swelling) with this shape change. This phenomena is called growth and its existence is now known to be not restricted to fissile metals.

In 1959 Cottrell proposed a model[4] to explain such growth. He recognised that such an orthorhombic crystalline material would have correspondingly anisotropic thermal expansion and that the heat generated locally by a fission spike would lead to a region of the crystal momentarily *contracting* in the 'b' direction and *expanding* in the 'a' direction. During this heat pulse he envisaged that extra atoms would flow *into* the regions between the hot and the surrounding cooler volume associated with the contraction directions and that there would be a compensating flow of atoms *from* the corresponding regions associated with the expansion direction. Needless to say he realised that, in a perfect crystal, such a process would be reversible on cooling so he invoked the presence of edge dislocations to act as appropriate sources and sinks for these atoms.

This homogeneous 'non-reversible' recrystallisation model of radiation growth is almost certainly incorrect but it did emphasise two important aspects that are now known to be important.

(i) Somehow there is a redistribution of atoms from the crystal planes orthogonal to the a-axis to form new atomic planes orthogonal to the b-axis.

(ii) Dislocations are important.

Rather than recrystallisation of the molten zone within the fission spike being important it is now known that the explanation lies with the intrinsic point-defects (interstitials and vacancies) generated by such spikes and their response to the thermal stress which is generated by the spike. Shortly after Cottrell's theory was published Hudson *et al.*[21]

showed by TEM that the growth in uranium has arisen from the presence of extensive sheets of interstitial loops on the [010] planes and vacancy loops on the [100] planes. Buckley[22] then showed theoretically that such preferred condensation of the two point-defect type aggregates should occur on precisely these different crystallographic planes, simply because of the intrinsic anisotropy of the thermal stresses arising from each fission spike. It has also been demonstrated[23] that the final arrangement into the sheets of loops can arise by the long-range elastic interactions between the loops inducing appropriate glide motion of the loops; this final re-arrangement is particularly assisted by the elastic anisotropy of α-uranium.[23]

It is interesting to speculate that the increase in growth due to cold work may well be due to a combination of local stress assisted vacancy emission from edge dislocations in regions under a tensile thermal stress with their corresponding loss at dislocations in regions under compression and also possible local stress-assisted preferred absorption (SIPA) processes which we will discuss in the context of radiation creep in the next sub-section.

Finally, since the time Cottrell considered the radiation growth of uranium, the phenomena of radiation growth has been observed in many non-fissile materials. The essential requirement is that the crystalline materials must be elastically anisotropic (non-cubic). Metals of great importance to the nuclear industry, such as zirconium and its alloys, have now been extensively studied. In essence, in such hexagonal metals the interstitials form dislocation loops on the prismatic planes and the vacancies either form dislocation loops on the basal planes or are lost at grain boundaries or at the network dislocations or are trapped at atomic trapping sites associated with large substitutional impurity atoms, such as Sn in Zircalloy etc. The rate theory of radiation damage, which will be alluded to below, has provided an accurate and quantitatively predictive model of the radiation growth process.[24, 25]

2.3.2 *Radiation creep of uranium and non-fissile materials*

Radiation creep is observed in most materials; it is a creep response to external stress that occurs at temperatures too low for thermal creep and only in the presence of radiation. Cottrell's explanation[4, 18] of this creep in uranium is particularly noteworthy since it is still accepted for uranium and has had considerable influence on our present understanding of the radiation creep of non-fissile materials.

Cottrell considered that the radiation creep observed in polycrystalline uranium was driven by the inexorable radiation growth (discussed in the previous section) in each grain. The impingement of these growing grains leads to the creation and maintenance of internal stresses in all the grains of the polycrystal. Under an external stress those grains that are stressed near yield will undergo plastic deformation in response to the external stress. The internal stresses will then relax until the inexorable growth returns the relevant grains to their plastic (near yield) state and another creep increment occurs and so on. This reasoning leads to the simple relation between the radiation creep rate $\dot{\epsilon}_c$ and the radiation growth rate $\dot{\epsilon}_g$

$$\dot{\epsilon}_c \approx [\sigma/\sigma_y]\dot{\epsilon}_g \tag{3}$$

where σ is the external stress. This expression remains consistent with much of the available data, although in the intervening years the switch to oxide fuels has meant that modern, more accurate data on the radiation creep of uranium has not been forthcoming.

Very many studies have been made on the radiation creep of non-fissile materials and it will suffice merely to comment on several important aspects of our current understanding.

(i) In polycrystalline materials that undergo radiation growth the internal stress mechanism postulated by Cottrell for uranium is clearly relevant.

(ii) Materials that do not grow (cubic materials etc.) also are observed to undergo radiation creep. For such materials we envisage that the external stress can itself induce a kind of growth process in each grain. Thus, the stress can enhance the probability of the planar interstitial aggregates nucleating on crystallographic planes that are orientated in sympathy (that is, orthogonal to a tensile stress axis) with the external stress.[26] The excess numbers of such interstitial loops in the 'aligned' orientation thus means that more extra lattice planes have been put down in this orientation orthogonal to the stress axis. The body has thus increased its length in the direction of the stress and radiation creep has occurred. Alternatively, the vacancies may also aggregate to form vacancy loops. In this case the stress can enhance the probability of the loops lying on suitable planes parallel to the stress. Again, such a process will enhance the length increase of the body in the direction of the tensile stress. This is the basis of the Stress Induced Preferred Nucleation (SIPN) mechanism of radiation creep. This mechanism predicts that once the nucleation regime is over, the creep rate ought to be independent of changes in the applied external stress; there is evidence that this does not occur and therefore we conclude that this mechanism cannot be appropriate at high radiation doses. Direct observation of such anisotropic loop distributions have been made in copper whiskers (see Ref. 2). Needless to say, it may be considered as a global form of the Cottrell mechanism of uranium growth where the internal thermal stresses from the fission spike are replaced by the external stress.

(iii) The presence of an external stress can modify the induced or inhomogeneity interaction between a dislocation and any nearby point-defects.[28, 29] In particular, it can be shown that edge dislocations should have an enhanced preference for interstitials when they are orientated with their extra 'plane' orthogonal to the tensile stress axis compared with the orthogonal orientations. This is the so-called Stress Induced Preferred Absorption (SIPA).[29] Its adoption leads to predicted radiation creep rates that are in good agreement with the data for steels with the great advantage of both conceptual and parametric simplicity.[30] When the dislocation density is very high it indicates, as is observed, radiation creep rates independent of the dislocation density and at low dislocation densities, such as the regime when SIPN might prevail, it will also be present; asymmetry in both the loop sizes and orientation would then be apparent.

(iv) In addition to the above two mechanisms, there are now known to be strong radiation creep effects arising from the preferred nucleation and growth of transmutation gas bubbles on grain boundaries orthogonal to the external stress axis. Such gas driven radiation creep[31] is important for 'in-pile' components and is usually far larger than the SIPA creep.

(v) The elongation associated with radiation creep can either arise from the actual climb of the dislocations or from the plastic flow induced by internal stresses generated by the climb processes. Rather sophisticated models which exploit the rate theory representation of radiation damage and the evolution of the entire microstructure now exist and can provide a quantitative description of the interplay between these alternatives.[32–34]

2.4 *Radiation swelling of materials*

The swelling observed in fissile and non-fissile materials when subjected to neutron irradiation is due to quite different origins, although in both cases we now know that the presence of rare gases produced by nuclear reactions is of crucial importance. When Cottrell studied the swelling of uranium in the mid-50s it was not suspected that non-fissile materials might also undergo a change of volume in addition to irradiation induced growth and creep. Cottrell's explanation[35, 36] for the swelling in uranium was simply that the insoluble rare fission gases such as krypton and xenon would aggregate and form gas bubbles. The difficult problem, at the time, was to explain the origin of the vacancies that must be provided to make space for the gas and to thereby maintain equilibrium between the gas pressure and the surface tension of the bubbles. In a rather beautiful paper Barnes, Redding and Cottrell[37] showed that the grain boundaries are the dominant source for the vacancies; this was proved by implanting helium atoms into copper then heating and observing the preferred growth of helium bubbles near the sources of vacancies — the grain boundaries.

This picture of swelling as being entirely gas-driven is almost certainly too simple, even for uranium. We now know that such swelling occurs in almost all non-fissile metals where there is nothing like enough helium, generated by n/α reactions, or other gases present to maintain equilibrium between the gas pressure and surface tension. That is the swelling is due to the formation and growth of cavities rather than gas bubbles! This phenomenon was first discovered at Dounreay in 1967[38] and led to a huge worldwide investigation, since its very existence threatened to jeopardise our ability to design a core for Fast Reactors that would be predictably stable. Needless to say, we now have a quantitative understanding of such swelling in many relevant structural materials and the designer can certainly cope with the situation.

Such swelling in non-fissile materials occurs because tiny gas bubbles nucleate by vacancy aggregation in conjunction with a few gas atoms (either helium generated by nuclear reactions or even trace gases that happen to be present in the material). If no gas is present we know that such vacancy aggregates will have a loop morphology and in fact many of the vacancy aggregates do undoubtedly form such loops owing to the absence of gas in the location of their nucleation. The important point is that such tiny gas bubbles can now start to grow into cavities with little or no further need for gas. We believe this happens* at sufficiently high temperatures because the interstitials and vacancies produced by the irradiation are then mobile and the interstitials, with their larger relaxation volume compared to vacancies, will be preferentially attracted to and lost at dislocations. It follows that because of this preferred loss of interstitials to dislocations there must be a concomitant net loss of vacancies to

* The possibility that this might happen was first suggested by Foreman, Greenwood and Rimmer[39] in a discussion of gas bubble nucleation and growth in fissile materials in 1959.

any *neutral* sinks in the material. The small gas bubbles are such relatively neutral sinks and therefore begin to grow by the net absorption of vacancies. Since there are dislocations present in uranium as well as in the non-fissile metals, there is no reason why such net vacancy flow to the gas bubbles should not also occur in such fissile materials.

The important point that should be emphasised in the present paper is that the fundamental basis for the phenomenon of swelling in non-fissile materials is the long-range interaction between the point-defects and dislocations and we have to thank Cottrell and Bilby[1] in 1949 for first investigating the form and strength of this interaction. It follows that, in addition to the specific contributions cited above, Cottrell may also be regarded as the father of the vast literature on microstructural and microchemical evolution during irradiation which has developed since the discovery of swelling in non-fissile materials in 1968.

3. Acknowledgement

It is a pleasure to thank Drs C. A. English, E. A. Little and W. J. Phythian for their advice and help with the preparation of this paper.

References

1. A. H. Cottrell and B. A. Bilby, *Proc. Phys. Soc. (London)* **62**, 49–62, 1949.
2. Lord Marshall of Goring, 'Second Report of the Study Group on the Integrity of Pressurized Light Water Reactor Pressure Vessels'. UKAEA, June 1982.
3. P. B. Hirsch, 'Addendum to the Second Report of the Study Group, since 1982.' UKAEA, April 1987.
4. A. H. Cottrell, in Proc. of the November Meeting of British Nucl. Energy Society, 'The 46th Thomas Hawksley Lecture', pp. 64–77, November 1959.
5. A. T. Churchman, I. M. Mogford and A. H. Cottrell, *Phil. Mag.* **2**, 1271–1275, 1957.
6. D. Hull and I. M. Mogford, *Phil. Mag.* **3**, 1213–1219, 1958.
7. E. A. Little, International Metal Reviews, **21**, 25–60, 1976.
8. N. Igata *et al.*, ASTM STP 529, 63–74, 1973.
9. E. A. Little and D. R. Harries, *Met. Sci. Jnl.* **4**, 188–199, 1970.
10. F. A. Smidt and J. A. Sprague, ASTM STP 529, 73–91, 1973.
11. W. J. Phythian and C. A. English, AERE Harwell Research Report, R-13632, 1989.
12. C. A. English, *J. Nucl. Mat.* **108/109**, 104–123, 1982.
13. T. J. Williams, P. R. Burch, C. A. English and P. H. N. de la cour Ray, 'The Effect of Irradiation Dose Rate and Temperature, and Copper and Nickel Content, on the Irradiation Shift of Low Alloy Steel Submerged Arc Welds,' in Proc. Third Int'l Symposium on Environmental Degradation of Materials

in Nuclear Power Systems — Water Reactors, NACE, TMS–AIME (Ed. George J Theus and John R Weeks), ANS, Traverse City, Michigan. Aug/Sept 1987, pp. 121–130. Published by Metallurgical Society, 1988.
14. T. J. Williams *et al.* in *Proc. 2nd International Symposium on Environmental Degradation of Materials in Nuclear Power Systems.* ASTM, Monteray, 1986.
15. H. Rauh and R. Bullough, *Proc. R. Soc. Lond.* **A397**, 121–141, 1985.
16. H. Rauh, C. A. Hippsley and R. Bullough, *Acta Metall.* **37**, 269–279, 1989.
17. A. J. Jacobs, R. E. Clausing, M. K. Miller and C. Shepherd, in *Proc. 4th International Symposium on Environmental Degradation of Materials in Nuclear Power Systems — Water Reactors,* Georgia, Nat. Assoc. of Corrosion Engineers, August, 1989.
18. A. C. Roberts and A. H. Cottrell, *Phil. Mag.* **1**, 711–718, 1956.
19. A. H. Cottrell, J. C. Bell, G. B. Greenough, W. M. Lomer and J. H. W. Simmons, in Proc. 2nd UN International Conf. on the Peaceful Uses of Atomic Energy, A/CONF.15/P/2485, August, 1958.
20. A. H. Cottrell, *J. Nucl. Mat.* **100**, 64–66, 1981.
21. B. Hudson, K. H. Westmacott and M. J. Makin, *Phil. Mag.* **7**, 377–392, 1962.
22. S. N. Buckley, in *Properties of Reactor Materials and the Effects of Radiation Damage* (ed. S. J. Littler). p. 413. London, Butterworths, 1962.
23. N. Meissner, Ph.D. Thesis of the University of Surrey, 1973, entitled 'General Anisotropic Continuum Green Function Representation with Applications to Defects in Copper and α-Uranium'.
24. S. N. Buckley, R. Bullough and M. R. Hayns, *J. Nucl. Mat.* **89**, 283–295, 1980.
25. R. Bullough and M. H. Wood, *J. Nucl. Mat.* **90**, 1–21, 1979.
26. A. D. Brailsford and R. Bullough, *Phil. Mag.* **27**, 49–64, 1973.
27. J. Y. Bonnaud, Thèse, 1973, Paris–Sud Orsay, reported by P. J. Leteurtre *et al.*, *Phil. Mag.* **27**, 1323–1334, 1973.
28. P. T. Heald and M. V. Speight, *Phil. Mag.* **29**, 1075–1080, 1974.
29. R. Bullough and J. R. Willis, *Phil. Mag.* **31**, 855–861, 1975.
30. R. Bullough and M. R. Hayns, *J. Nucl. Mat.* **57**, 348–352, 1975.
31. R. Bullough, D. R. Harries and M. R. Hayns, *J. Nucl. Mat.* **88**, 312–314, 1980.
32. F. R. N. Nabarro, R. Bullough and J. R. Matthews, *Acta Metall.* **30**, 1761–1768, 1982.
33. R. Bullough and M. W. Finnis, in *The Mechanics of Dislocations* (eds E. C. Aifantis and J. P. Hirth). p. 11. Ohio, Am. Soc. of Metals, 1983.
34. J. R. Matthews and M. W. Finnis, *J. Nucl. Mat.* **159**, 257–285, 1988.
35. A. H. Cottrell, *Met. Rev.* **1**, 479–522, 1956.
36. A. T. Churchman, R. S. Barnes and A. H. Cottrell, *J. Nucl. Energy* **7**, 88–96, 1958.
37. R. S. Barnes, G. B. Redding and A. H. Cottrell, *Phil. Mag.* **3**, 97–99, 1958.
38. C. Cawthorne and E. J. Fulton, *Nature* **216**, 575–576, 1967.
39. A. J. E. Foreman, G. W. Greenwood and D. E. Rimmer, *J. Nucl. Mat.* **4**, 305–324, 1959.

7 'The Metallic State — Revisited'

P. P. Edwards

Lecturer, Department of Chemistry, University of Cambridge

Presented at the Meeting 'Advances in Physical Metallurgy' to mark the 70th birthday year of Sir Alan Cottrell.

Abstract

It is no exaggeration to say[1] that, of all materials in common use, none has had a greater influence on our technological development — and indeed our existence — than Metals. Their mechanical strength, ductility, high electrical and thermal conductivities, and considerable alloying abilities qualify them to occupy a pivotal role in the physical sciences for which there is no substitute.

However, in 1967, R. S. Nyholm[2] commented that, with the notable exception of Linus Pauling,[3] the subject of Metals and the metallic state was much neglected by Chemists. Fortunately, this is no longer the case. The last 30 years have seen an increasing awareness by Chemists of the correlation between structure and properties of the metallic state[4–6] and an unprecedented interest in the subject of metal–metal bonding.[7]

My own interest in Metals and Insulators, and the Transition between them, was triggered partly by an article[8] published in 1960 by A. H. Cottrell, entitled 'The Metallic State'. This perceptive article, and other major contributions by N. F. Mott[9] and L. Pauling,[3] illustrated that an atomic, or 'chemical' description was extremely useful when it came to thinking about the differences between Metals and Insulators.[10]

Sir Alan's seminal work in physical metallurgy will form the basis of other contributions to this volume. My own presentation is to acknowledge his major influence on my own views and ideas of Metals and Insulators. Some 30 years on from Sir Alan's 1960 article,[8] I would like to discuss . . . 'The Metallic State — Revisited'.

1. What, Why and When is a Metal?

For several hundred years, man has sought to classify elements and materials in terms of their metallic or insulating properties. Despite a long awareness of this division in kind, a fundamental understanding of the nature of, and differences between, Metals and Insulators has only slowly begun to emerge.

As Cottrell has recently pointed out:[10] 'The question ... 'What is a Metal?' has a long and surprising history of answers.' Before delving into this perplexing and philosophical question,[11] we should stress that a given element or material is neither inexorably a Metal nor inevitably an Insulator. On the planet Jupiter, for example, where ambient conditions are cold and dense, most elements, including hydrogen, are believed to be metallic.[10] From this, one can conclude that it is possible to induce a transition from an Insulator to a Metal, or vice-versa, by varying the 'conditions of ambience' — *intensive* thermodynamic conditions such as temperature, pressure, density and/or composition.[4]

It is entirely reasonable to expect a sharp distinction in nature between Metals and Insulators for temperatures approaching the absolute zero.[11] A Metal would exhibit a finite, or infinite* value of the electrical conductivity, $\sigma(0)$ at $T = 0K$, while an Insulator would have zero conductivity at $T = 0K$. This obvious, but fundamental, distinction is illustrated quite dramatically by 'zero temperature' conductivity studies by Rosenbaum and co workers on silicon doped with phosphorus[13] (Fig. 1). To determine $\sigma(0)$, data are taken at as low a temperature as possible (usually a few millikelvin) and then extrapolated to $T = 0K$.

It can then be seen the Insulator-to-Metal Transition in P:Si occurs at a critical phosphorus (donor) density (n_c) of 3×10^{18} electrons cm^{-3}.[14] For a sample with a donor density (n) less than n_c, the material is clearly an Insulator for $T \rightarrow 0K$. For $n > n_c$ the material is a Metal ($\sigma(o)$ finite as $T \rightarrow 0K$). These experiments beautifully illustrate the controlled genesis of a Metal from an Insulator.

At considerably higher temperatures, it is always possible, at least in principle and often in practice, to transform, for example, an *elemental* metal into an insulating state via expansion to low densities.[15] Thus the alkali metals, those traditional paradigms of the metallic state, may be transformed continuously from a metallic to an insulating state by gradually reducing the density of the system.[16] By expanding a metallic liquid (in this case caesium) up to supercritical conditions (1750 °C and 114 bar), and thereby reducing the number density, a transition from a metallic to a non-metallic state may be continuously followed (Fig. 2).

It is important to note that both the high (Fig. 2) and low (Fig. 1) temperature systems, at all densities, would be characterised by a half-

* In the superconducting state.

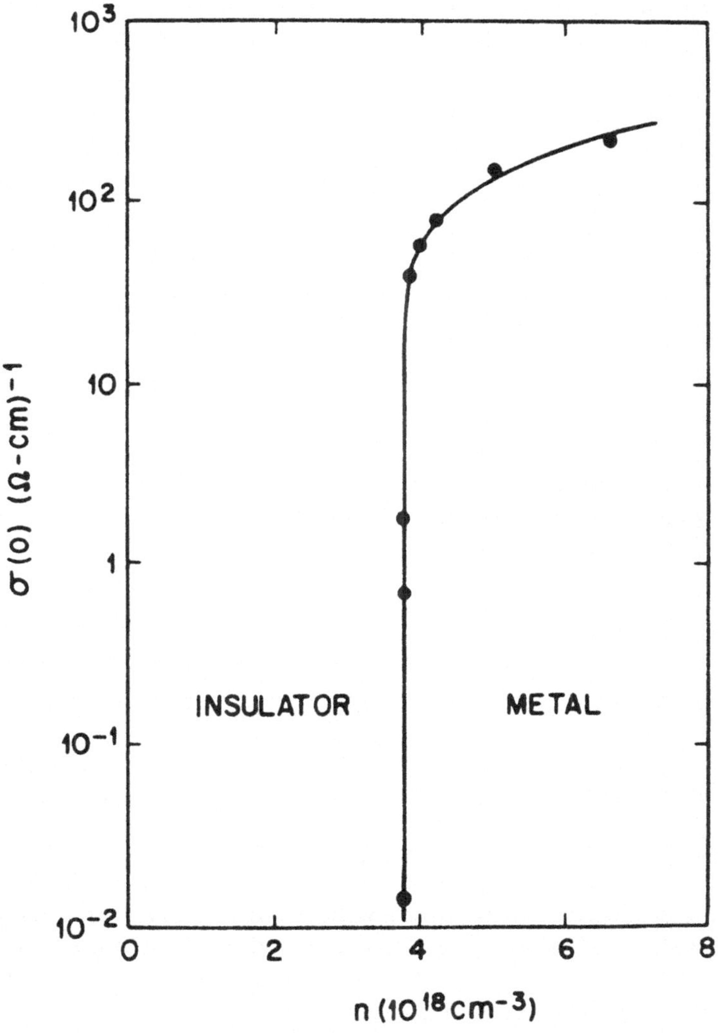

Fig. 1. A semilog plot of zero-temperature conductivity, σ (o), versus donor density, n, *for Si:P samples. The transition is extremely sharp, but apparently continuous.*

Taken from Rosenbaum *et al.*,[13] and used with permission.

filled conduction band. Without the inclusion of electron correlation and disorder, a classical band model would therefore fail to predict a Metal–Insulator Transition at any density. As Mott pointed out in his celebrated paper[9] of 1961, '. . . this is against common experience and, one might say, common sense'!

One cannot hope, therefore, to understand the nature and condition

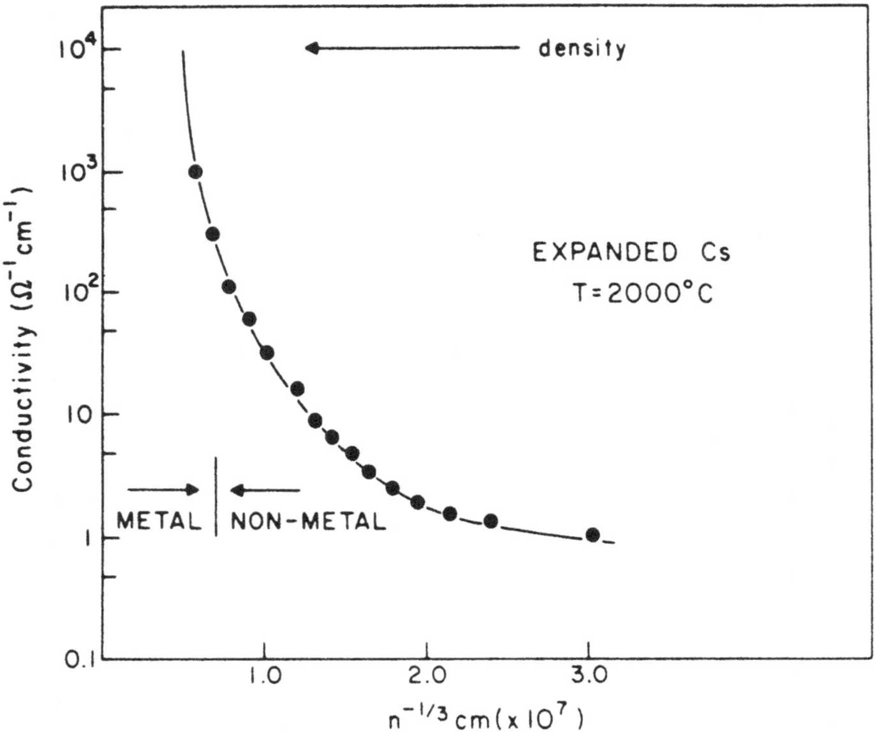

Fig. 2. Electrical conductivity of elemental caesium across the Metal–Insulator Transition. The concentration of caesium is given as the cube root of the reciprocal number (electron) density.
Taken from Hensel,[16] and used with permission.

of the metallic or insulating (nonmetallic) states of matter in isolation from an understanding of the Transition between these two canonical states of matter: The two questions *'What is a Metal?'* and *'When does a Metal transform to an Insulator?'* are inseparable (see Section 3).

The problem can be approached in a slightly different manner; namely the concept of physical 'size' and the metallic state.[17–21] To understand either the metallic or insulating regimes one must also elucidate the manner in which these (bulk) macroscopic states of matter evolve from the progressive agglomeration of their respective atomic constituents. Two sodium atoms, for example, will form a neutral Na_2 molecule; three, four and so on will form larger molecular clusters, and the process continued *ad infinitum* leads to a macroscopic structure — bulk sodium — which exhibits metallic behaviour under normal conditions. However, there is clearly a fundamental and qualitative difference between a piece of bulk, metallic sodium and a relatively small collection of sodium atoms, despite the fact that both consist of the same element.[18] This leads us to the question of the size dependence of the structural, electronic and dynamical properties of isolated clusters or particulates, consisting of a few — up to several hundred — atoms or molecules. The interest here is to probe the electronic level structure and dynamics in going from a single atom to a bulk metal.[18, 20] Thus, any Metal–Insulator Transition in such a *single* microscopic particle must arise by changing the *extensive* thermodynamic properties of the state (total mass, volume and absolute particle number). Hence the question *'How Many Atoms Maketh Metal?'* (see Section 4).

How, therefore, can one understand the difference between Metals and Insulators? This question was posed by Cottrell[8] in 1960 in the article 'The Metallic State'. At that stage it was gradually emerging that there were at least three approaches to the problem.[10, 22, 23] The older one (deriving from the work of Sommerfeld, Bloch and Wilson) assumed the existence of free, or itinerant electrons and then set out to prove under certain circumstances the electrons can carry a current and under the other conditions they cannot.[22] Mott, in a series of pioneering studies, pointed out that this approach, although elegant and comprehensive in situations *where it works*, can also predict to be Metals some substances that are known to be Insulators. Cottrell pointed out[8] 30 years ago 'this is . . . not quite the right criterion for deciding between Metals and Insulators'!

A second approach, still in its infancy in the early 60s, starts with the view that valence electrons are localised in distinct, spatial chemical bonds in Insulators, and considers the conditions under which these valence electrons become ionised into mobile, free carriers and a concomitant 'metallic orbital'. This particular theory of the difference between Metals and Insulators was due mainly to Mott,[9] but was closely related to the theory of Pauling,[3] involving resonating valence bonds (RVB). Interestingly, the RVB model, proposed by Pauling to understand metallic valence and the structure of metals, has also recently been invoked[24] for the occurrence of high-temperature superconductivity in oxides.

A third approach, by Anderson,[25] was concerned with the effects of disorder, or randomness, in producing electron localisation or itinerancy. Anderson showed that even without electron interactions, localised electronic states can result from the presence of a random potential sufficiently large compared to the (one-electron) bandwidth. The relative importance of electron correlation and disorder effects near the Metal–Insulator Transition has been, and still is, hotly debated.[26]

Another approach, developed by Herzfeld in 1927, was based on the notion of a Dielectric Catastrophe at the Insulator-to-Metal Transition.[27] For a variety of undoubtedly complex reasons, the Herzfeld view of

Metals and Insulators lay dormant for many years. Recently, however, there has been a renaissance of interest in the approach.

The aim of my contribution, then, is to briefly review and elaborate on certain developments in how we might think about Metals and Insulators.

2. Diversity of the Metallic State: Electronic Structure and Classification of Materials

The enormous range of electrical conductivities exhibited by elements and materials is a remarkable fact. Between the most conductive and most resistive, the difference in electrical conductivity can amount to something like 10^{23}. Of course, in the present day, the class of materials which we label 'Metals' is now very much larger than ever envisaged.[10] It goes well beyond the traditional elemental metals of the Periodic System, and now encompasses oxides, sulphides, selenides, etc, and even organic materials, viz. chemical compounds as well as chemical elements.[4] Clearly one must now also take into account the phenomenon of high transition-temperature superconductivity in metallic oxides.[28] The recent breathtaking advances in this field also reopens the major question of the origin of metallic character in oxides and related materials. As Sir Alan has put it[10] '. . . the best conductor of electricity at the temperature of liquid nitrogen is now no longer a traditional metal, but an oxide. It is possible that the same may one day be true at room temperature.'

These observations, and the great interest in identifying new materials and new phenomena close to the Metal–Insulator Transition, are compelling reasons for developing more powerful classifications of the metallic, nonmetallic and other states of matter.[29–31]

There have been a number of phenomenological approaches to classifying solids;[32, 33] generally these are based on the type of chemical bond which is responsible for the binding energy of materials. Seitz's early classification[33] contains 5 categories: (1) Metals (2) Ionic Crystals (3) Valence or Covalent Crystals (4) Molecular Crystals and (5) Semiconductors. In certain respects, this type of classification is entirely artificial since, as we shall illustrate, certain of these modes of chemical bonding can be entirely transformed via changes in the relevant thermodynamic quantities. More recent efforts to systematise and characterise materials[29–32] rely strongly on realistic descriptions of electronic structure — *and how this electronic structure may be manipulated* — in relation to conductivity, magnetism and so on.

Specifically, the important concept of phase boundaries in the classification scheme is now introduced; such boundaries[29–31] separate qualitatively distinct regimes of electronic structure. In its broadest sense, these schemes propose that all condensed systems can be considered as belonging to essentially three canonical states of matter which, in their simple limits, are (1) Metals (2) Paramagnetic Insulators (sometimes called Mott Insulators) and (3) Diamagnetic Insulators (sometimes called Covalent Insulators). One such derivative of modern schemes[29–31] (illustrated by a triangular plot) shows the three limiting canonical bonding types separated by phase boundaries (Fig. 3). Here the characteristic bonding energy (and therefore the chemical bonding in the solid) is mainly determined by the interplay of electron delocalization, electron correlation and electron-lattice (vibronic) effects. These three interactions are characterised, respectively, by the relevant energy contributions T, U and V.

Within such a scheme[29–31] — in which attempts are made to determine

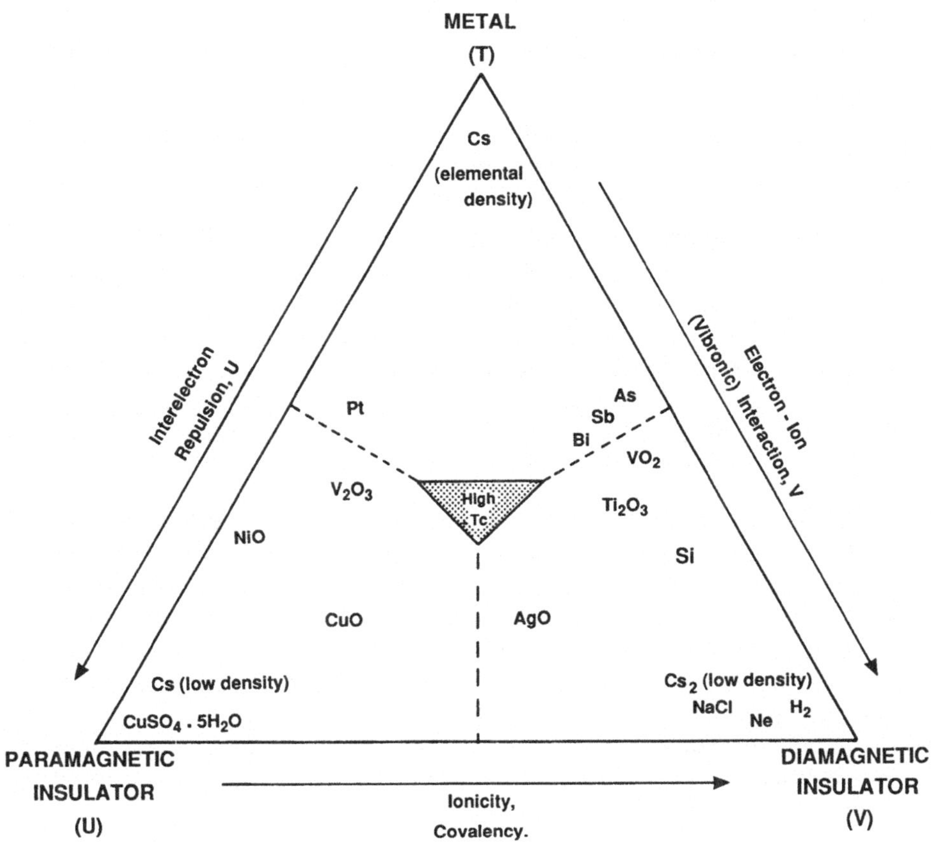

Fig. 3. The three limits of condensed matter adapted from Chakraverty,[20] Wooley[30] and Urland,[31] and used with permission.

and/or predict T, U and V in a self consistent fashion — these parameters form the basis of a broad classification of materials. Furthermore, there are 3 major limiting cases which can give rise to 3 canonical electronic structures:

$T > U, V$ 'Metals'
$U > V, T$ 'Paramagnetic Insulators'
$V > T, U$ 'Diamagnetic Insulators'.

These 3 states of matter show clearly the Metal–Insulator phase boundary to be expected as we move away from the simple Metal limit ($T > U,V$) on the U, or V, axes. Another axis, the U–V line, links the Paramagnetic and Diamagnetic Insulators, and have an appropriate ionicity or covalency parameter encompassed within U,V. It is relatively easy to define a simple Metal in such a description; here the Kinetic Energy of the itinerant electron gas dominates all electronic interactions, such that $T > U,V$. As the Kinetic Energy (one-electron band width) becomes smaller two distinct, and different, symptoms may signal the transition from Metal to Insulator.[29, 30] These are:

(i) an enhanced magnetic susceptibility (and specific heat) which indicate strong repulsive interactions between itinerant electrons (viz, an enhanced U term).
(ii) an enhanced vibronic interaction (enhanced V) leading to a reduced density-of-electronic states ($N(E_F)$), electron-paring, diamagnetism and (potential) superconductivity. More specifically, a non-zero V has two strikingly related effects, both related to V-induced attractive interaction between two electrons. For $V<<T$, this results in electron-pairing and superconductivity while for $V>>T$, we find localised pair ground states which are diamagnetic singlet states.

Materials with particularly interesting electronic properties are found on these various phase boundaries[29, 30] (Fig. 3). Urland[31] has proposed a cursory classification of conventional and oxide superconductors as materials located near the 'Triple-Point' boundaries of Fig. 3. Thus, the occurrence of superconductivity is clearly associated (at this qualitative level) with both the localization (U) and vibronic (V) interactions in barely-metallic materials ($T \sim U \sim V$).

Finally, to return to the important notion of phase boundaries separating distinct regions of electronic structure.[29, 30] Crossing such a boundary leads to a change in bond type and to:

(i) Qualitative changes in electronic structure; Metal–Insulator Transition, Diamagnet–Paramagnet Transition etc.

(ii) Concomitant discontinuities (often genuine phase transitions) in magnetic electrical and other properties.

Of course, it is important when considering such diagrams to reiterate that a Metal–Insulator Transition can take place even for simple, one-component Systems, e.g. the alkali metals. As we have shown earlier, by expanding the liquid alkali metal up to supercritical conditions, a Metal–Insulator Transition may be continuously followed with no accompanying liquid-gas transition. This conceptually simple system of a lattice of H-like atoms emphasises the all-important relationship between density and the metallic state. In the case of expanded caesium (Fig. 2) there exists a fundamental, and qualitative difference between the properties of low-density (and potentially metallic) caesium and those of the high density form. As the density of elemental caesium is reduced — effectively reducing T while increasing U and possibly V (Fig. 3) — a transition to the nonmetallic form of caesium will inevitably take place. In the following section we outline an old and, until recently, half forgotten pre-quantum theory which attempts to understand this important feature; *Density and the Metallic State.*[27, 34]

3. Density and the Metallic State

Dielectric catastrophe: origin and observation

Almost certainly the first attempts to explain the occurrence of Metals and Insulators were made by Goldhammer[34] in 1911, and Herzfeld[27] in 1927; the latter in a paper entitled 'On Atomic Properties which make an Element a Metal'. In simple terms the Goldhammer/Herzfeld approach is concerned with the mutual polarisation of an atom by the remaining atoms in the condensed phase (liquid or solid). If this interaction is large, a polarization catastrophe occurs, the valence electrons acquire itinerancy and the material becomes a Metal.[23]

This approach is based on what is known as a polarisation or dielectric catastrophe and the idea is that the dielectric constant of the material should diverge to infinity as the Insulator–Metal Transition is *approached from the insulating side*. A simple description of the dielectric behaviour of a non-polar insulator is embodied in the familiar Clausius–Mossotti (CM) relation which, for a one-component system, takes the form

$$\frac{\epsilon - 1}{\epsilon + 2} = \frac{4\pi}{3} n\alpha_o \qquad (1)$$

where α_o is the static (scalar) polarisation of an isolated atom and n is the (number) density of atoms. The essence of the Herzfeld metallisation criterion is that as $n \rightarrow n_c$ (the critical density); $n_c = 3/(4\pi\alpha_o)$, $\epsilon \rightarrow \infty$ and a second-order dielectric constant is predicted. To relate explicitly the polarisation catastrophe, as predicted by the CM relation, with the onset of electron itinerancy, Herzfeld[27] considered a simple mechanical model (the Drude oscillator model) used in the *classical* theory of electrons; here the valence electrons of an atom are regarded as harmonically bound to their parent atoms.

Consider an isolated atom with a single valence electron of mass m in an external electrostatic field, $\mathbf{E}^o$, such that

$$\mathbf{p} = e\mathbf{r} \tag{2}$$

be the instantaneous dipole associated with the instantaneous displacement, $\mathbf{r}$, of the harmonically bound electron from its parent atom. The classical equation of motion for $\mathbf{p}$ is then

$$m\ddot{\mathbf{p}} + m\omega_o^2\,\mathbf{p} = e^2\mathbf{E}^o \tag{3}$$

where ω_o is the characteristic angular frequency of the valence electron.

If now this atom is immersed in a condensed medium, $\mathbf{E}^o$ must be replaced by the effective field acting to polarise the atoms in this medium, which can be taken to be the Lorentz local field

$$\mathbf{E}^{LO} = \mathbf{E} + (4\pi/3)n\langle\mathbf{p}\rangle \tag{4}$$

where $\langle\mathbf{p}\rangle$ is the mean dipole moment of an atom. The equation of motion for $\mathbf{p}$ is then

$$m\ddot{\mathbf{p}} = e^2\mathbf{E} - m\omega_o^2 p + (4\pi/3)ne^2\langle\mathbf{p}\rangle \tag{5}$$

where $(m/e)\ddot{\mathbf{p}} = \mathbf{F}$ is the instantaneous force acting on the valence electron. As the valence electron is bound to its parent atom (in Insulators, at least) the mean force $\langle\mathbf{F}\rangle$ acting on it must be zero, then

$$\langle\mathbf{p}\rangle = \frac{\alpha_o}{1 - (4\pi/3)n\alpha_o}\mathbf{E} \tag{6}$$

where $\alpha_o = e^2/(m\omega_o^2)$ is the polarisability of the harmonically bound electron. From (6) $\langle\mathbf{p}\rangle \rightarrow \infty$ as $n \rightarrow n_c = 3/(4\pi\alpha_o)$ and the dielectric catastrophe occurring as $n \rightarrow n_c$ in the CM equation, is associated with the onset of electron itinerancy, manifest in the resultant mechanical instability of the harmonically bound electron. The CM relation is commonly cast in the form

$$\frac{\epsilon - 1}{\epsilon + 2} = R/V \tag{7}$$

where V is the molar volume and R is the long wavelength limit of the (low density) molar refractivity,

$$R = 4\pi N_A \alpha_o/3 \tag{8}$$

(N_A is the Avogadro number).

The Herzfeld criterion,[27] then, predicts that a material will show metallic behaviour (and the associated dielectric catastrophe) if the molar volume is reduced to, or below, the critical value $4\pi N_A \alpha_o/3$. This simple criterion, requiring solely a knowledge of the free atom polarisability, was used by Herzfeld[27] in 1927 to explain why elements in the Periodic System are metallic, insulating or borderline when in the condensed state.

Just how well the overall features of the Periodic System of the Elements conform to the Herzfeld criterion can be seen by referring to Fig. 4 which shows[35] the data for the naturally occurring elements at their ambient (normal atmospheric) densities. For example, the Group VIIA elements (the halogens) and the rare gases all have $R/V < 1$ and nonmetallic behaviour is predicted, and observed, for these elements. Conversely, the Group IA elements (the alkali metals) all have $R/V > 1$ and metallic character. The elements of Groups IIIA–VIA effectively straddle the dividing line between localised and itinerant electron behaviour.

There has recently been considerable interest in observing directly the onset of an impending dielectric catastrophe in systems which transform from insulating to metallic behaviour via changes in composition, pressure, etc. Castner *et al.*[36] have measured the dielectric constant for n-doped (Sb, P, As) silicon and germanium at temperatures down to 11 mK. Figure 5 shows the results for Sb, P and As-doped Si samples as a function of $n(\equiv N_D - N_A)$, where N_D and N_A are the concentration of donors and acceptors, respectively. ϵ is the appropriate dielectric constant of the 'impurity band' found by subtracting the host value ($\epsilon_{Si} = 11{\cdot}4$) from the measured dielectric constants. As $n \rightarrow n_c$ (n_c is in the region of 3×10^{18} elec. cm.$^{-3}$), there is clearly a steep rise in ϵ signifying the onset of a polarisation catastrophe in the transition to the metallic state.

In terms of the donor (or impurity) polarisability in the doped semiconductors, α_D, the Clausius–Mossotti relation gives

$$\epsilon = \frac{\left(1 + 8\pi \frac{n\alpha_D}{3}\right)}{\left(1 - \frac{4\pi n\alpha_D}{3}\right)} \tag{9}$$

A polarization catastrophe is predicted when $(4\pi n\alpha_D/3) = 1$, which occurs at a critical concentration n_c, given by $n_c^{1/3} a_H \sim 0{\cdot}38$,[35, 36] where a_H is the Bohr radius of the isolated impurity centre.

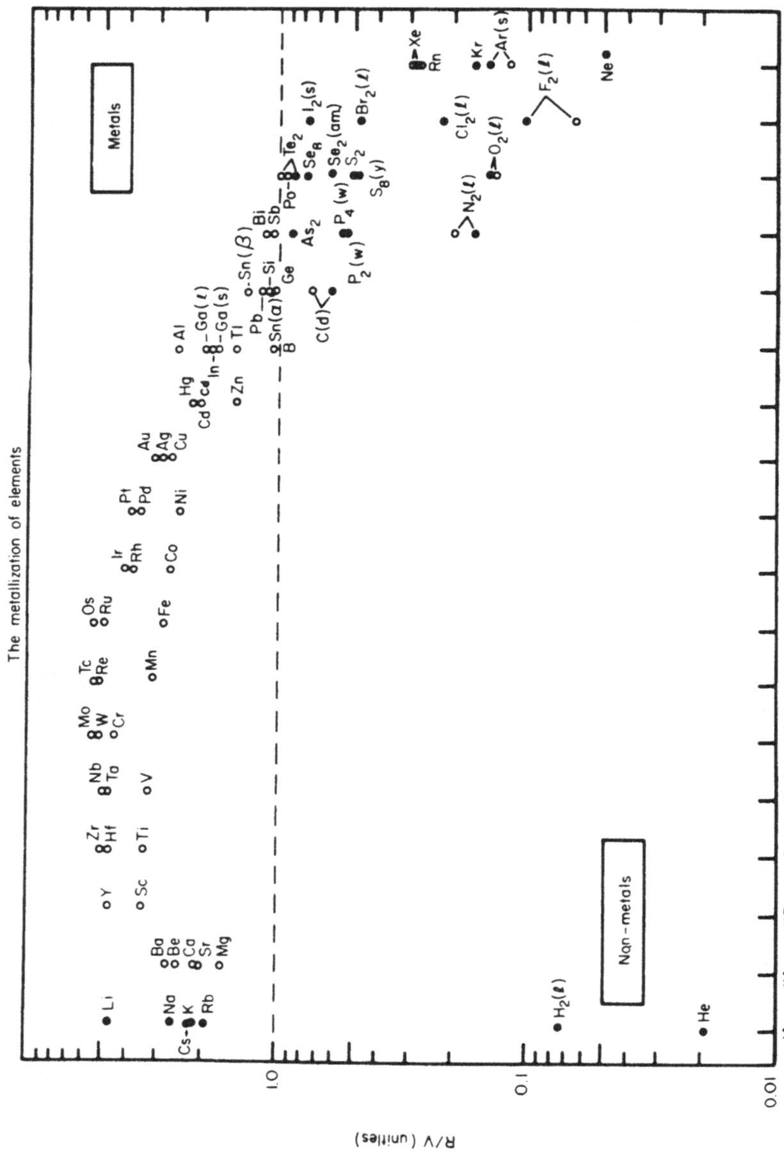

Fig. 4. The metallisation of elements. The figure shows the ratio R/V for elements of the s, p and d blocks of the Periodic Table.

Taken from Edwards and Sienko.[35]

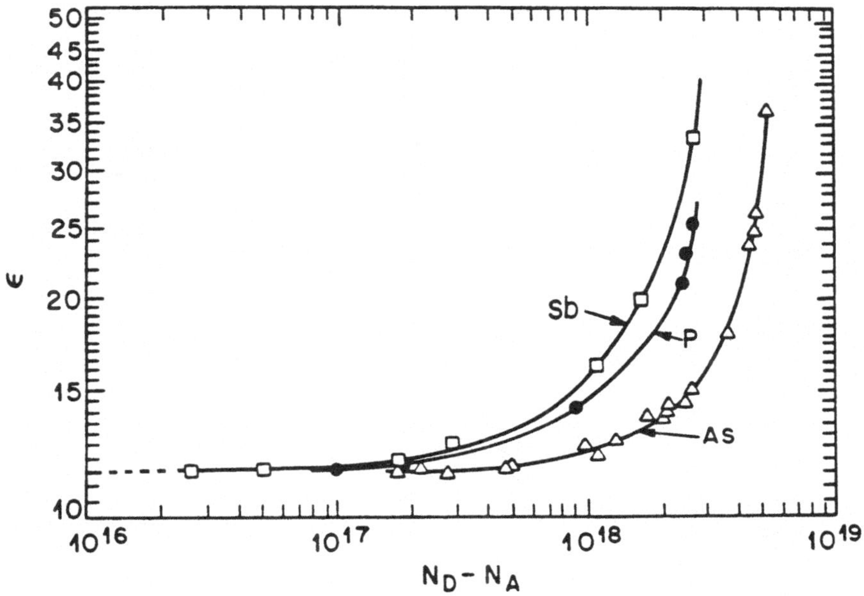

Fig. 5. The dielectric constant of (Sb,P,As):Si (millikelvin temperatures) as a function of the donor concentration, expressed as N_D–N_A *(see text for details).* Taken from Castner,[36] and used with permission.

Another method of obtaining ϵ, and the associated dielectric susceptibility, χ, is by integrating the far IR absorption coefficient using the Kramers–Kronig relationship.[39] The susceptibility is related to the dielectric constant, ϵ, via $\epsilon = \epsilon_{Si} + 4\pi\chi$.

Figure 6 shows the results[39, 40] for P-doped Si at millikelvin temperatures, in which $4\pi\chi$ is plotted as a function of n, the P-donor concentration. The corresponding electrical conductivity, extrapolated to $T = 0K$, ($\sigma(o)$), is also shown. Clearly, giant dielectric constants characterise the approach to the metallic state in this system where the Metal–Insulator Transition can be conveniently 'tuned' by the amount of P doping into crystalline Si. Similar results (although, in certain cases, at considerably higher temperatures) have also been reported for amorphous Au:Si alloys, n-type Ge, Hg–Xe thin films and (fluid) metal-ammonia solutions (see refs. 11, 35). In summary, the experimental data for a wide range of systems lend credance to the Herzfeld idea of an impending polarisation catastrophe at the critical density for metallisation.

The Herzfeld criterion predicts that a material will show metallic behaviour if the molar volume (V) is reduced below the molar refractivity (R, obtained at standard conditions). This criterion, at varying levels of complexity and sophistication[23] has been applied to systems exhibiting

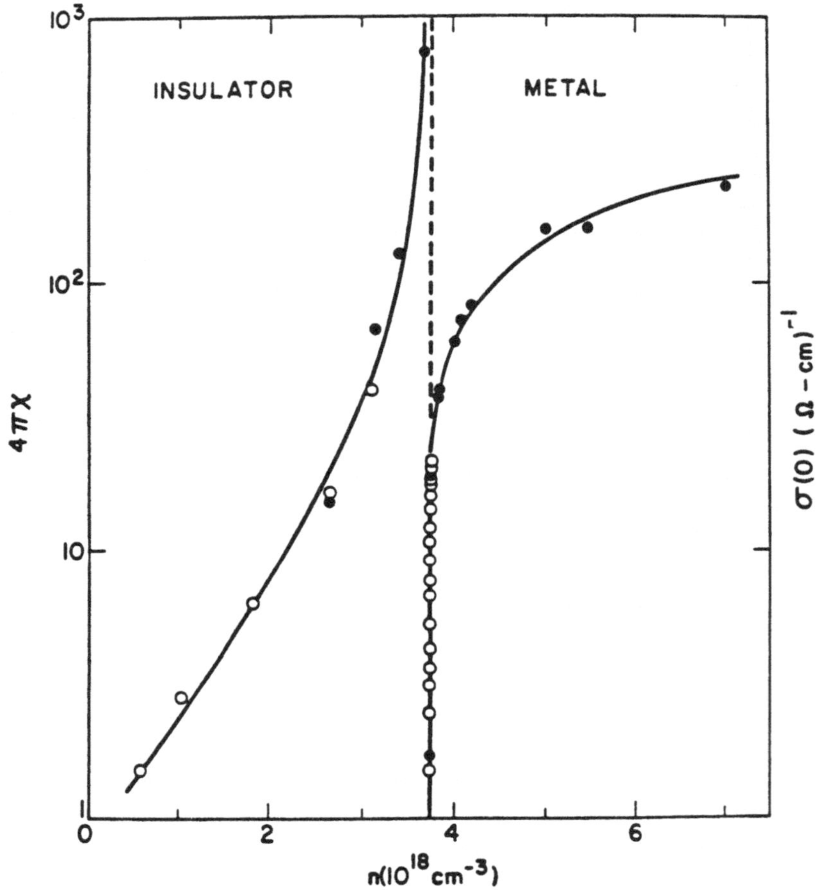

Fig. 6. Giant dielectric constants at the approach to the Insulator–Metal Transition. The figure shows the divergence of the (T = 0K) *donor (phosphorus) dielectric susceptibility* ($4\pi\chi$) *in the insulating regime, and the corresponding* (T = 0K) *conductivity as a function of donor density.*

Taken from Hess *et al.*[40] and used with permission.

induced Metal–Insulator Transitions in order to predict the critical experimental conditions for metallisation. In particular, the criterion has found widespread applicability in the realm of high-pressure experiments. Two recent, important experimental advances are outlined here.

Metallisation of xenon and hydrogen

In 1989 there have been two reports[41, 42] of the direct observation of the transition to metallic xenon (at 150 GPA pressure) as well as optical evidence for the band-overlap mechanism of metallisation in hydrogen, at pressures above 200 GPa.

We see from Fig. 4 that Xe has the lowest predicted metallisation pressure of all the noble gases. Herzfeld,[27] in 1927, predicted that Xenon would go metallic at a density of 10·2 cm^3/mole; this is the condition under which the molar volume of Xe would become equal to its (gas-phase) molar refractivity. In 1989, the experimentally determined[41, 42] molar volume of Xe at metallisation is 10·70(12) cm^3/mole, only slightly larger than the original Herzfeld prediction. Electron band-structure calculations have shown that the transition of Xe to the metallic state at high pressure is due to a crossing of the empty 5p valence bands.[43]

Goettel *et al.*[42] have investigated the metallisation of Xe by obtaining optical data in both the metallic and insulating states. In the metallic state, the pressure dependence of the plasma frequency (ω_p) was determined from absorption data fitted to a Drude, free-electron model.

In the insulating state, the pressure dependence of the band gap was determined from absorption data fitted to an indirect band gap model. At the critical metallisation density, the band gap goes to zero. These authors have interpreted both the visible and infrared optical data for Xe to obtain three independent values of the Insulator-to-Metal transition density; combining these three determinations they observe metallisation at

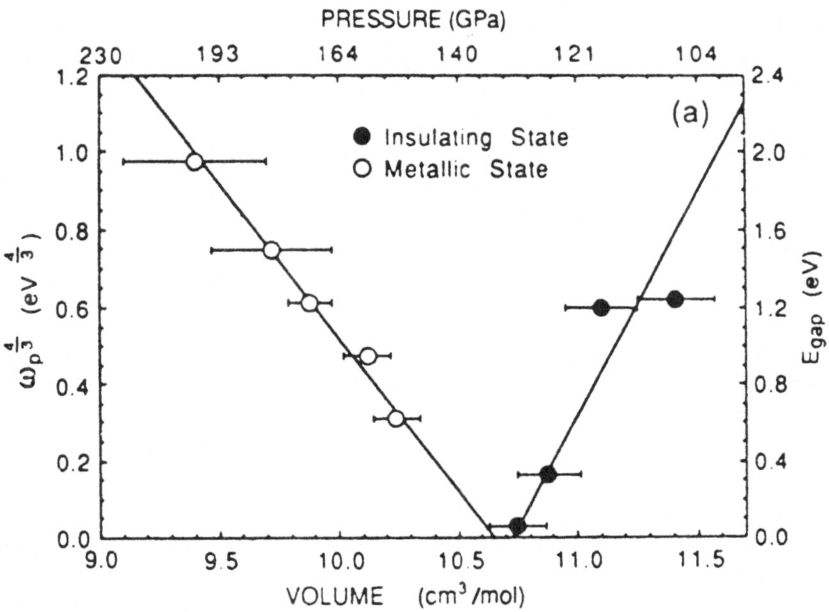

Fig. 7. The metallisation of xenon.
Taken from Ref. 42 and used with permission.

10.70(12) cm^3/mol, or 132(5)GPa (Fig. 7). Reichlin *et al.*[41] have monitored the transformation of Xe to a metallic state near 150 GPa, obtaining absorption and reflection data across the Insulator–Metal Transition.

One of the most sought-after experimental results in the area of pressure induced phase transitions is the conversion of molecular hydrogen to metallic hydrogen. Metallic H_2 has, for a long time, enjoyed the favour of a great many theoreticians since, in certain of its nonmetallic forms (H, H_2), it represents a substance whose properties may be calculated from first principles. Yet, paradoxically, it offers enormous resistance to experimental investigation, primarily because of the unearthly pressures required for its metallisation. One proposal is that with increasing pressure, H_2 will first become a conducting molecular solid, then an atomic metallic solid.[44, 45] As with Xe, one assumes that the Insulator–Metal Transition in hydrogen takes place when the valence-conduction energy band gap closes.

It is generally believed[46] that the best approach to the metallisation of hydrogen is by means of the diamond anvil cell where static pressures of up to 550 GPA have been obtained.

The Herzfeld criterion gives a predicted metallisation pressure for hydrogen of between 270 and 280 GPa, corresponding to a metallisation volume of 2·02 cm^3/mole. Mao and Hemley[47] have recently reported direct optical absorption of solid H_2 to pressures in the 250 GPa range at 77 K. Hydrogen samples become opaque at the maximum pressures. These authors demonstrate,[47] by the use of Raman spectroscopy, that the molecular solid is stable to at least 200 GPa and 77 K. The electronic excitation in the visible region of the spectrum begins in the 200 GPa range and the optical data can be interpreted in terms of a decreasing band gap to form a high pressure metallic, but apparently molecular, solid. Figure 8 summarises the experimental and theoretical information[47] on the volume (pressure) dependence of the band gap and absorption threshold in hydrogen.

The Raman effect observed at 180 GPa (2·40 cm^3/mol) have been interpreted as tracking the band edge (~2·5 eV) close to the metallisation onset. Interestingly, these pressures are in the range expected for dissociation of the molecular bond as well as the simple Herzfeld model predictions for metallisation. The characterisation of hydrogen at these extreme pressures represents an exciting challenge not only for high-pressure technology, but also for other areas of condensed matter science.[48] For example, in 1968 Ashcroft published an article[49] predicting that metallic hydrogen might be a potential candidate for high temperature superconductivity — possibly even room temperature!

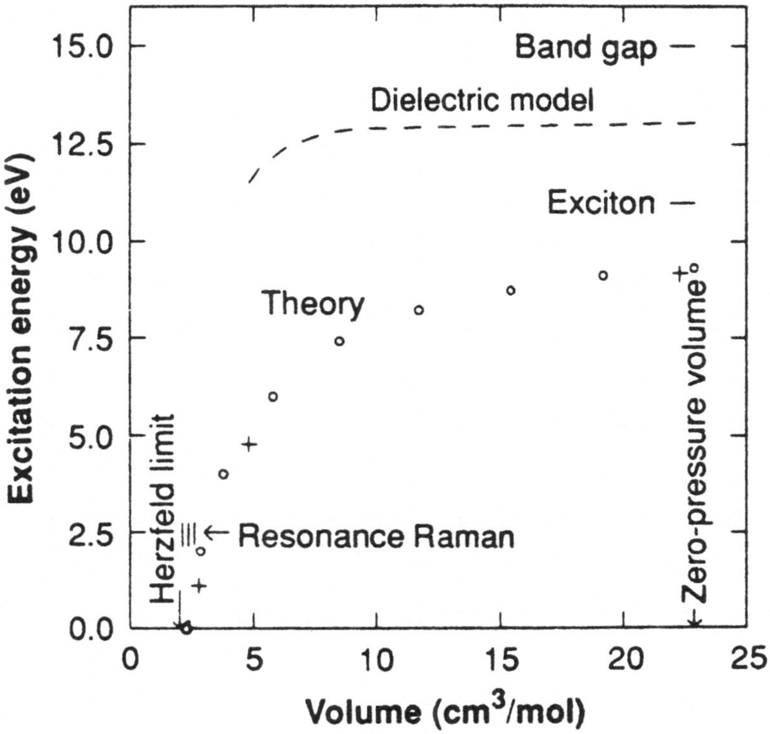

Fig. 8. The approach to metallisation in hydrogen. The plot shows the volume dependence of the band gap in solid hydrogen. The Herzfeld limit (see text) is indicated, as well as the volume dependence of optical properties, etc. Also included is the zero-pressure volume.

Taken from Mao and Hemley,[47] and used with permission.

Metals on Uranus

The Herzfeld criterion has also found application in 'planetary' considerations. Torbett and Smoluchowski[50] have explained the presence of a magnetic field on Uranus using a three-layer model of the planet. The core of Uranus is metallic, mostly liquid and gravitational differentiation produces enough heat in it to drive a hydromagnetic dynamo.

Specifically, the planetary core is considered[51] to consist of a mixture of 38, 25, 25 and 12% (by mass) of MgO, SiO_2, FeS and FeO, respectively. Figure 9 shows the three layer model of Uranus according to Hubbard and McFarlane.[51] The model consists of a 'rocky' core, an 'ice' mantle, and a H_2–He envelope comprising 25, 65 and 10% of the planet by mass, respectively. The existence of a metallic, conducting envelope for H_2 and

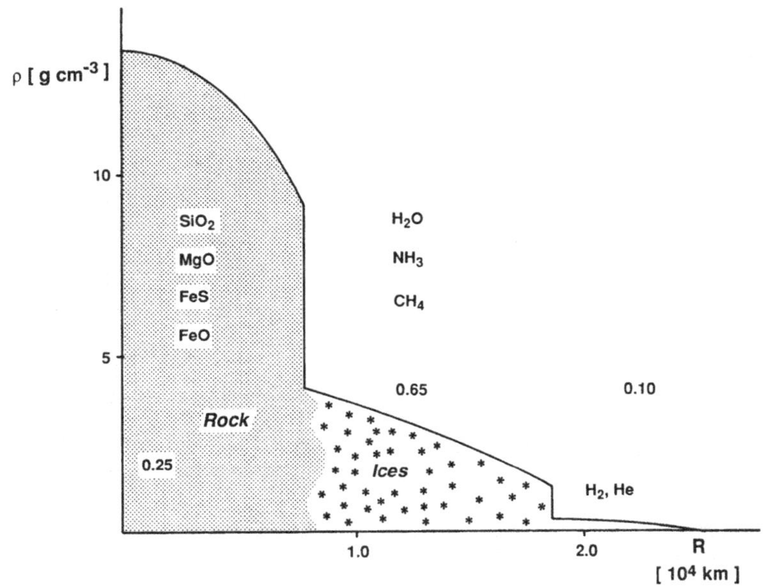

Fig. 9. A three-layer model of Uranus according to Hubbard and MacFarlane. Taken from Torbett and Smoluchowski,[51] and used with permission.

He is ruled out since the planetary pressure is well below the predicted metallisation value for these elements.

To investigate the electrical conductivity of the core, Torbett and Smoluchowski[50, 52, 53] applied Herzfeld's criterion[27] for each of the core oxides and sulphides. The results are shown in Figure 10 where metallic conductivity is indicated by solid lines. MgO and SiO_2 are predicted to be metallic up to about 85% of the radius of the core, FeO is metallic throughout the core and FeS is known to be metallic even in the kilobar range of pressures. The authors conclude,[50, 52, 53] therefore, that the bulk of the core possesses metallic conductivity, and the existence of a magnetic field on Uranus can be understood as being a natural result of gravitational differentiation in this liquid, metallic core.

A description of this 'old' (pre-quantum) description of Metals and Insulators has been advanced by Cottrell[10] . . . 'A material becomes a metal when its atoms are brought sufficiently close together for the energy cost of transferring a valence electron from one atom to another . . . is . . . more

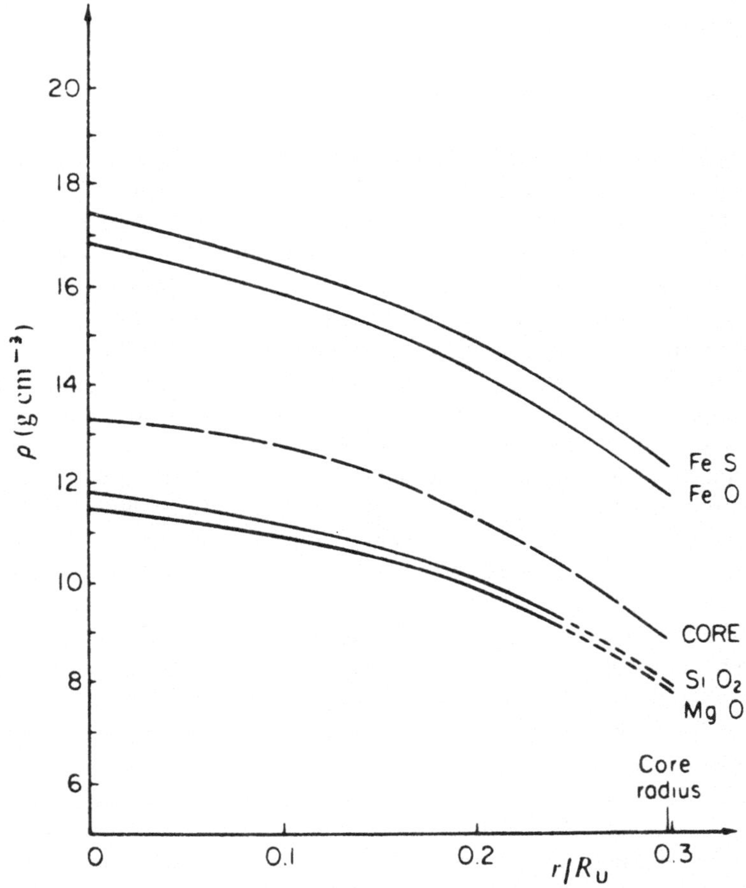

Fig. 10. The radial dependence of the core density of Uranus, and the density of its constituents. The full line indicates metallic properties, the dashed line indicates nonmetallic behaviour (as gauged from the Herzfeld criterion). Taken from Torbett and Smoluchowski,[53] and used with permission.

than offset by the energy that would be gained by "*liberating*" the electron from its atomically localised state into the free electron state'.

This chemically-appealing approach,[27] outlined over 60 years ago, is one way of understanding, and predicting, the conditions under which valence electrons are ionised from their atomic constituents to yield the metallic state. It now seems that the question 'What is a Metal?' can now be recast 'When is a Metal?'

4. Size and the Metallic State

(1) How Many Atoms Maketh Metal?

The preceding descriptions have focused on condensed phases of macroscopically extended crystalline and amorphous systems. Here, variations in *intensive* thermodynamic quantities (temperature, density, pressure and molar composition) produce a transition between metallic and insulating states. However, a Metal–Insulator Transition *must* also occur by decreasing the *extensive* thermodynamic variables of state (e.g. total mass, volume and absolute particle number) of a *single cluster or aggregate of atoms*.[54] In this particular area of interest, there are two closely interwoven problems.[17–21] The first is the inevitable Metal–Insulator Transition in an individual grain of Metal as its dimensions are continuously reduced to the atomic limit. The second is the question of the nature and electronic structure of small — but still metallic — particles.[18–21] A small metallic particle can be considered as a solid with geometric dimensions comparable to, or smaller than, some intrinsic length scale, e.g. the wavelength of light, the electron mean-free path,[18] etc.

We focus here primarily on the question of the Metal–Insulator Transition in a cluster of atoms; the structure, electronic properties and dynamics of small metallic particles are outlined elsewhere.[18–21]

The inevitable consequence of the successive division of a piece of metal must be the cessation of 'metallic' character below some critical particle size. It is interesting to note that Mott,[9] in 1961, made reference to such a situation. Whereas he proposed a first-order (discontinuous) Metal–Insulator Transition in macroscopic systems, Mott noted[7] that '. . . the sharp transition described here is only expected in an infinite lattice. It goes without saying that for a *finite* number of atoms there will be a gradual decrease in the weight of ionised states . . . as the interatomic distance is increased or in other words, a gradual transition.'

There seems to be agreement that this transition will be continuous through some size range, but considerable disagreement on its width and the location of the appropriate size range.[54] Thus, Perdey[55] proposes that even a single sodium atom should be labelled 'metallic', while Quinn *et al.* argue[56] that several millions of atoms in a cluster are needed to reach bulk, metallic properties. Wood and Ashcroft[19] propose that the Metal–Insulator Transition in an isolated particle may be viewed in terms of a (modified) Ioffe–Regel localisation criterion.[26] This criterion, originally proposed for macroscopic systems, signals the onset of the breakdown of the Boltzmann equation for conduction electron transport, viz:

$$k_F l \leqslant 2\pi$$

where l is the electron mean free path and k_F is the Fermi wave vector $(=\lambda^{-1})$ appropriate to the electron density. Ioffe and Regel[26] argued that since l is the distance over which the electron wavefunction loses phase coherence, and the electron wavelength $(\lambda \sim k_F{}^{-1})$ is the distance over which it changes phase by 2π, it does not make sense to have $\lambda >> l$ for metallic character.

Wood and Ashcroft[19] suggest that if one crudely interprets l as a characteristic small particle dimension (say, the particle diameter, d) then the cessation of conducting behaviour in a small particle may be interpreted as a Metal–Insulator Transition of the Ioffe–Regel sort. For most metals, the transition would therefore occur for particle diameters below *ca.* 10–15 Å.

During the past two decades, there have been major developments in the study of both small metallic particles and the associated Metal–Insulator Transition. We concentrate here on an important study of the Metal–Insulator Transition in isolated atomic clusters of mercury.[54]

(2) *Case Study: The Metal–Insulator Transition in Mercury Atom Clusters*

Rademann *et al.* have recently investigated[52] the electronic structure of isolated Hg clusters up to a mass range of approximately 40,000 amu. These authors find a gradual transition from van der Waals type bonding to metallic cohesion occurs in the size range between 20 and 70 atoms. Moreover, photoelectron spectra[54] indicate that the evolution of the valence band structure toward the (bulk) metallic type (6s6p) is almost fully accomplished for isolated mercury clusters larger than Hg_{70}.

A plot of the measured ionization potentials for Hg_x as a function of cluster size is shown in Fig. 11. By assuming spherical symmetry, the ionisation potential of a cluster can be expressed in terms of a size-dependent work function, W(d), given by

$$W(d) = W_\infty + \frac{e^2}{d}$$

where d is the diameter of the equivalent sphere supporting an elementary charge and W_∞ is the (bulk) metal work function.

For bulk mercury (liquid or solid), the work function is 4·49 eV. This will be increased by $e^2/_d$ (=7·2 eV/Å) as long as the droplet (or cluster) indeed remains metallic in the sense that the electrons are delocalised and have a finite density-of-states ($N(E_F)$) at the Fermi level (Region I). Beginning at the bulk value, this term is plotted as a broken line in Fig. 11.

When the sphere becomes smaller and smaller (i.e. decreasing num-

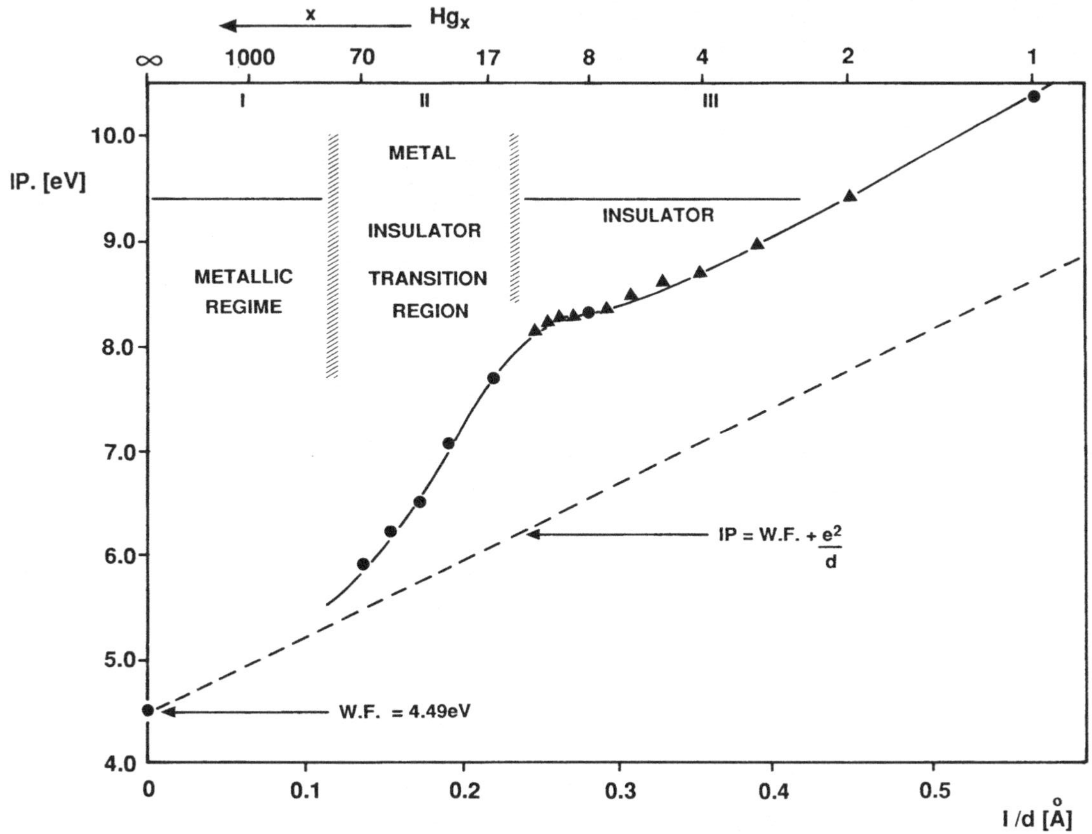

Fig. 11. The variation in the measured ionisation potention as a function of cluster size for isolated clusters of Hg atoms. The work function (4·49 eV) for bulk Hg is indicated. The dashed line is a plot of the classical electrostatic formula for a metallic sphere of diameter d. *Region III contains clusters with fully occupied s-states derived from atomic s-orbitals. Region II denotes the Metal–Insulator Transition region, in which overlap of the 6s and 6p states sets in at (approximately)* Hg_{13}. *Larger cluster, Region I, have a valence electronic level structure that closely resembles the band structure of liquid and crystalline mercury.*

Taken from Rademann[54] and used with permission.

ber of atoms), a departure from the straight line signals the appearance of so-called 'Quantum Size Effects'.[18, 20] These effects result from the finiteness of the sample and the concomitant discrete nature of the electronic energy levels.

For a particle containing N atoms, an average energy level spacing, δ, is given by[19, 54]

$$\delta = 4E_F/3N$$

where E_F is the Fermi energy of the bulk metal. In this approximation, the average electronic energy level spacing is thus:

0·009 eV for Hg_{1000}
0·09 eV for Hg_{100}.

If this level spacing becomes comparable to, or larger than, some characteristic energy scale, e.g. thermal energy (kT), quantum-size effects should be observed.

The appearance of a significant increase in the ionisation potential for clusters smaller than Hg_{70} (Fig. 11) is proposed to be a symptom of such effects. This also signals incipient electron localisation for 'nonmetallic' clusters smaller than this size. Approaching from the nonmetallic side (Region III), we note that the transition region is characterised by a large drop in the ionisation potentials of *ca.* 2eV between Hg_{13} and Hg_{70}.

It is interesting to note that these results[54] for isolated Hg atom clusters are consistent with related effects found for the Metal–Insulator Transition in expanded fluid mercury.[15, 16] In this 'macroscopic' system, a Metal–Insulator Transition occurs in the density range 8–9 g/cm^3. Rademann[52] has established a proportionality between the density of expanded fluid mercury and an effective coordination number. $\bar{N}_{eff}$. If $\bar{N}_{eff}$ becomes less than *ca.* 5 the system will be in an essentially nonconducting state. For $\bar{N}_{eff} \lesssim 6$–8, the expanded fluid exhibits metallic properties.[15, 16]

In the case of the isolated Hg clusters, the Metal–Insulator Transition region (II in Fig. 11) is reflected in a large drop in the average coordination number, from

to $\bar{N}_{eff} = 6$ at Hg_{16}
$\bar{N}_{eff} = 5$ at Hg_{10}.

Thus, the gradual *size-dependent* Metal–Insulator Transition in isolated Hg clusters is proposed to be induced by the decrease in the average number of nearest neighbour Hg atoms in the cluster.

5. Concluding Remarks

In this short contribution, I have attempted to illustrate some of the recent advances in terms of describing, and probing, the metallic state. The coverage has been representative and certainly not exhaustive. One feature that I have tried to illustrate is the development of our view of the Metal–Insulator Transition and the old — but important — discussion of the atomic properties which make materials metallic.

Finally, in this 70th birthday presentation to Sir Alan, it is perhaps worthwhile to briefly mention the origin[57] of the word 'Metal'. This derives from the Greek μεταλλαω (metallō) . . . 'to search out', and this seems an appropriate description of elemental metals. However, there is a further derivative,[57] Metallizomai, meaning . . . 'to be condemned to hard labour in the mines'. This, perhaps, may be our fate until we meet again in 10 years' time to celebrate Sir Alan's 80th birthday!

6. Acknowledgements

I would like to thank Sir Alan Cottrell and Dr David Logan for many interesting and stimulating discussions. I gratefully acknowledge the B.P. Venture Research Unit for their continued support and encouragement of my work.

References

1. *The Cambridge Guide to the Material World*, R. Cotterill, Cambridge University Press, Cambridge, 1985.
2. R. S. Nyholm, Presidential Address to the British Association, in *Advancement of Science*, p. 421, 1967.
3. L. Pauling, *Phys. Rev.* **B54**, 899 (1938); *Proc. R. Soc.* **A196**, 343, 1949; for a recent article, see *J. Solid State Chem.* **54**, 297, 1984.
4. See, for example, articles in *The Metallic and Nonmetallic States of Matter*, Eds. P. P. Edwards and C. N. R. Rao, Taylor and Francis, London, 1985.
5. A. Simon, *Angew. Chem. Int. Ed. Engl.*, **27**, 159, 1988.
6. R. Hoffmann, *Solids and Surfaces: A Chemist's View of Bonding in Extended Structures*, V.C.H. Publishers, New York, 1988.
7. *Advanced Inorganic Chemistry*. F. A. Cotton and G. Wilkinson, John Wiley & Sons, New York, 1986.
8. A. H. Cottrell, *Contemporary Physic* **1**, 417, 1960.
9. N. F. Mott, *Proc. Phys. Soc. Lond.* **62**, 416, 1949; *Nuovo Cim. Suppl.* **7**, 318, 1958; *Phil. Mag.* **6**, 287, 1961.

10. *Introduction to the Modern Theory of Metals*. A. H. Cottrell, The Institute of Metals, London, 1988.
11. P. P. Edwards and M. J. Sienko, *Int. Rev. Phys. Chem.* **3**, 83, 1983.
12. W. D. Metz, *Science*, **183**, 398, 1973.
13. T. F. Rosenbaum, R. F. Milligan, M. A. Paalaneu, G. A. Thomas, R. N. Bhatt and W. Liu, *Phys. Rev.* **B27**, 7509, 1983; and T. F. Rosenbaum, K. Andres, G. A. Thomas and R. N. Bhatt, *Phys. Rev. Lett.* **45**, 1723, 1980; T. F. Rosenbaum, K. Andres, G. A. Thomas and P. A. Lee, *Phys. Rev. Lett.* **46**, 568, 1981.
14. R. F. Milligan, in Ref. 4.
15. F. Hensel, in Ref. 4.
16. F. Hensel, *Ber. Bunsenges, Phys. Chem.* **75**, 619, 1971; *Angew. Chem., Int. Ed. Engl.* **19**, 593, 1980; in NATO ASI Series C (Eds, J. V. Acrivos, N. F. Mott and A. D. Yoffe). (Dordrecht: Reidel) **30**, 401, 1984.
17. P. P. Edwards, in 'Advances in Solid State Chemistry', Ed. C. N. R. Rao, Proceedings of the Indian National Science Academy Golden Jubilee Symposium, INSC, New Delhi, 1986, p. 265.
18. J. A. A. J. Perenboom, P. Wyder and F. Meier, *Phys. Report* **78**, 173, 1981.
19. D. M. Wood and N. W. Ashcroft, *Phys. Rev.***B25**, 6255, 1982.
20. W. E. Halperin, *Rev. Mod. Phys.* **58**, 535, 1986.
21. *Physics and Chemistry of Small Clusters* (Eds P. Jena, B. K. Rao and S. N. Khanna). Plenum Press, New York, 1987.
22. 'The Beginnings of Solid State Physics', *Proc. Roy. Soc.* **371**, 1980.
23. D. E. Logan and P. P. Edwards, in Ref. 4.
24. P. W. Anderson, *Science*, **235**, 1196, 1987.
25. P. W. Anderson, *Phys. Rev.* **109**, 1492, 1958.
26. *Metal–Insulator Transitions*, N. F. Mott. Taylor and Francis Ltd, London, 1988.
27. K. F. Herzfeld, *Phys. Rev.* **29**, 701, 1927.
28. K. A. Müller and G. Bednorz, *Proc. Natl. Acad. Sci. U.S.A.* **84**, 4678, 1987.
29. B. K. Chakraverty, *Nature* **287**, 393, 1980.
30. R. G. Wooley, in *Molecules in Physics, Chemistry and Biology, Vol. II* (ed. J. Marvani), Kluwev Academic Publishers, p. 651, 1988.
31. W. Urland, *Angew. Chem. Int. Ed. Engl.* **26**, 1275, 1987.
32. P. W. Anderson, *Concepts in Solids*. W. A. Benjamin, Inc., 1965.
33. F. Seitz, *Modern Theory of Solids*. McGraw Hill, New York, 1940.
34. D. A. Goldhammer, *Dispersion und Absorption des Lichtes*. Teubner, Leipzig, 1911.
35. P. P. Edwards and M. J. Sienko, *Acc. Chem. Res.* **15**, 87, 1982.
36. T. G. Castner, *Phil. Mag.* **B42**, 873, 1980; T. G. Castner, N. K. Lee, G. S. Cieloszyk and G. L. Salinger, *Phys. Rev. Lett.* **34**, 1627, 1975; T. G. Castner, N. K. Lee, H. S. Tau, L. Moberly and O. Symko, *J. Low Temp. Phys.* **38**, 447, 1980.
37. K. F. Berggren, *J. Chem. Phys.* **60**, 3399, 1974.
38. H. Fritzsche, in *The Metal–Nonmetal Transition in Disordered Systems* (ed. L. R. Friedman and D. Tunstall). NATO ASI, SUSSP Publications, Edinburgh, p. 193, 1978.
39. M. Capizzi, G. A. Thomas, F. De Rosa, R. N. Bhatt and T. M. Rice, *Phys. Rev. Lett.* **44**, 1019, 1980.

40. H. F. Hess, K. De Conde, T. F. Rosenbaum and G. A. Thomas, *Phys. Rev.* **B25**, 5578, 1982.
41. R. Reichlin, K. E. Brister, A. K. McMahan, M. Ross, S. Martin, Y. K. Vohra and A. L. Ruoff, *Phys. Rev. Lett.* **62**, 669, 1989.
42. K. A. Goetel, J. H. Eggert, I. F. Silvera and W. C. Moss, *Phys. Rev. Lett.* **62**, 665, 1989.
43. M. Ross, *J. Chem. Phys.* **56**, 4651, 1972.
44. E. Wigner and H. B. Huntingdon, *J. Chem. Phys.* **3**, 764, 1935.
45. L. G. Caron, *Comments Sol. State Phys.* **6**, 103, 1975.
46. I. F. Silvera, private communication.
47. H. K. Mao and R. J. Hemley, *Science,* **244**, 1462, 1989.
48. I. F. Silvera, preprint, 'The Phase Diagram and Excitations in Solid Hydrogen: Prospects for Metallization'.
49. N. W. Ashcroft, *Phys. Rev. Lett.* **21**, 1748, 1968.
50. M. Torbett and R. Smoluchowski, *Nature,* **286**, 237, 1980.
51. W. B. Hubbard and J. J. MacFarlane, *J. Geophys. Res.* **85**, 225, 1979.
52. M. Torbett and R. Smoluchowski, *Geophys. Res. Letts.* **6**, 675, 1979.
53. M. Torbett and R. Smoluchowski, *Proc. 11th Lunar Planet Sci. Conf.* Pergamon Press, p. 1907, 1980.
54. K. Rademann, B. Kaiser, U. Even and F. Hensel, *Phys. Rev. Lett.* **59**, 2319, 1987; K. Rademann, *Ber Bunsenges. Phys. Chem.* **93**, 653, 1989.
55. J. P. Perdew, *Phys. Rev.* **B37**, 6175, 1988.
56. C. M. Quinn, J. G. R. McKiernan and D. B. Redmond, *J. Chem. Soc.* Faraday Trans. 2, **80**, 1507, 1984.
57. I. M. Le M. Du Quesnay, private communication.

8 Microstructural Characterisation of Defects and Mechanisms (including field–ion observations)

B. Ralph

Professor, Department of Materials Technology, Brunel University of West London

Abstract
A brief personal overview is presented on this topic aimed at demonstrating the synergism between theoretical developments and microscopical mechanisms in physical metallurgy/materials science. It is shown that many developments in microscopical techniques have as their origin a need for higher resolution in order that defects may be characterised.

1. Introduction

Microscopy has played a key element in the scientific development of physical metallurgy as an academic discipline and in establishing the basis of materials science. Much of the theoretical development of this subject arises from experimental results obtained by modern microscopical techniques. Perhaps an attraction of our subject is the close relationship which has developed between theoreticians and experimentalists. In turn, because of the need in the materials field for ever more refined microscopical results, materials scientists have been at the vanguard of developing microscopical techniques and the interpretation of the results which they present.

From the times of people like Sorby and Heycock, microscopy has been of crucial importance from both scientific and technological standpoints in establishing the grain and phase populations in engineering materials. On the one hand we use microscopy to establish the importance of grain size in technological terms and on the other our appreciation of hardening mechanisms due to precipitation, dispersions and fibres owes much to microscopical studies. At the end of the day, the result is 'designed microstructures' which have been produced to optimise a 'basket' of properties (so far mainly mechanical properties) such as in superalloys and composites. This interplay of science and technology is of crucial importance to key technologies such as the aerospace and energy industries. Much of the microscopical activity now centres on non-metallic materials with properties other than mechanical (e.g. magnetic, semiconducting and superconducting) being the key areas.

In one other area, that is, education, the importance of microscopy cannot be overstressed. Here, microscopy helps to form the scientific base from which students in this subject are taught and avoids the 'Mrs Beeton's cookery book' approach. Whilst microscopy initially had (and obviously continues to have) an impact on the physical side, more and more microscopical observations are used to underpin our knowledge and the teaching of the chemical and processing sides of this subject. It is perhaps germain here to mention that Alan Cottrell introduced a generation of Cambridge undergraduates to chemical metallurgy through a masterful mechanistic approach.

Perhaps a part of the art of being great is to be at the right place at the right time and to realise that this is the case. Alan Cottrell's arrival at Cambridge was at a key time in the development of the techniques of electron microscopy (by the teams lead by Jack Nutting and Peter Hirsch). In particular, the theory of diffraction contrast was in its infancy and just getting to the point where the characterisation of dislocations and other defects became possible. Further, Erwin Müller had just demonstrated atomic resolution in the field–ion microscopy and Alan Cottrell was instrumental in setting up a team primarily concerned with the application of this instrument to the examination of defects and mechanisms in metals and alloys.[1]

2. The Search for Atomic Resolution

It is easy to overstress the importance of the current generation of microscopic (e.g. electron and field–ion) techniques and forget the rôle that light microscopy has played and the importance it still has. Further whilst scanning electron microscopy is associated with lower resolution than modern transmission electron microscopes, in practical terms, it represents a much more important investigative technique for establishing mechanisms of failure (e.g. Fig. 1) and as the basis of a technique for 'writing' submicrometre electronic circuits.

One aspect of the scanning electron microscope is that it presents not only a series of electron images conveying different aspects of information about the structure (topographical and atomic number) but that a whole series of other signals are available (current, cathodeluminescent

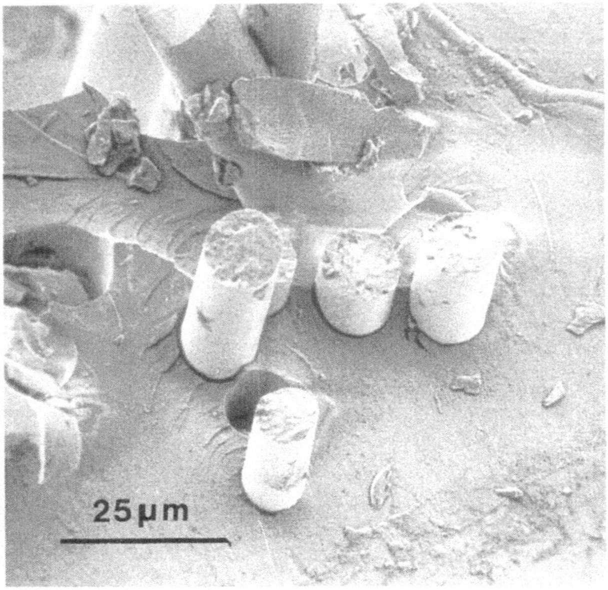

Fig. 1. Scanning electron micrograph of a fracture surface in a unidirectional glass fibre reinforced polyester composite (from F. Guild and B. Ralph, J. Mat. Sci. **14**, *2555–2562, 1979.*

and X-ray) which give information on the chemical distribution of species, local crystallography, etc.). Accordingly modern theoretical developments tends to incorporate chemical factors as well as pure geometry, for instance in the treatment of grain boundary embrittlement. In this particular case, that is embrittlement, a vast array of modern instruments is used, many of these are primarily associated with surface science (e.g. secondary ion mass spectroscopy, X-ray photo-electron spectroscopy and scanning Auger microscopy (e.g. ref. 2) and it becomes, to say the least, hard to define what is a microscope (e.g. ref. 3). A further example of the problem of limiting the breadth of the definition of what is a microscope would be the case of the scanning acoustic microscope; in essence one form of non-destructive testing brought up-to-date and permitting the resolution of larger scale defects and mechanisms in bulk samples.

The rapid development in dislocation theory in the decade or so immediately after the Second World War 'spurred on' the development of techniques towards every higher resolution. Lattice imaging of large lattice parameter crystal structures (e.g. platinum pthalocyanine) lead to almost direct imaging of individual dislocations.[4] Eventually, using

(a)

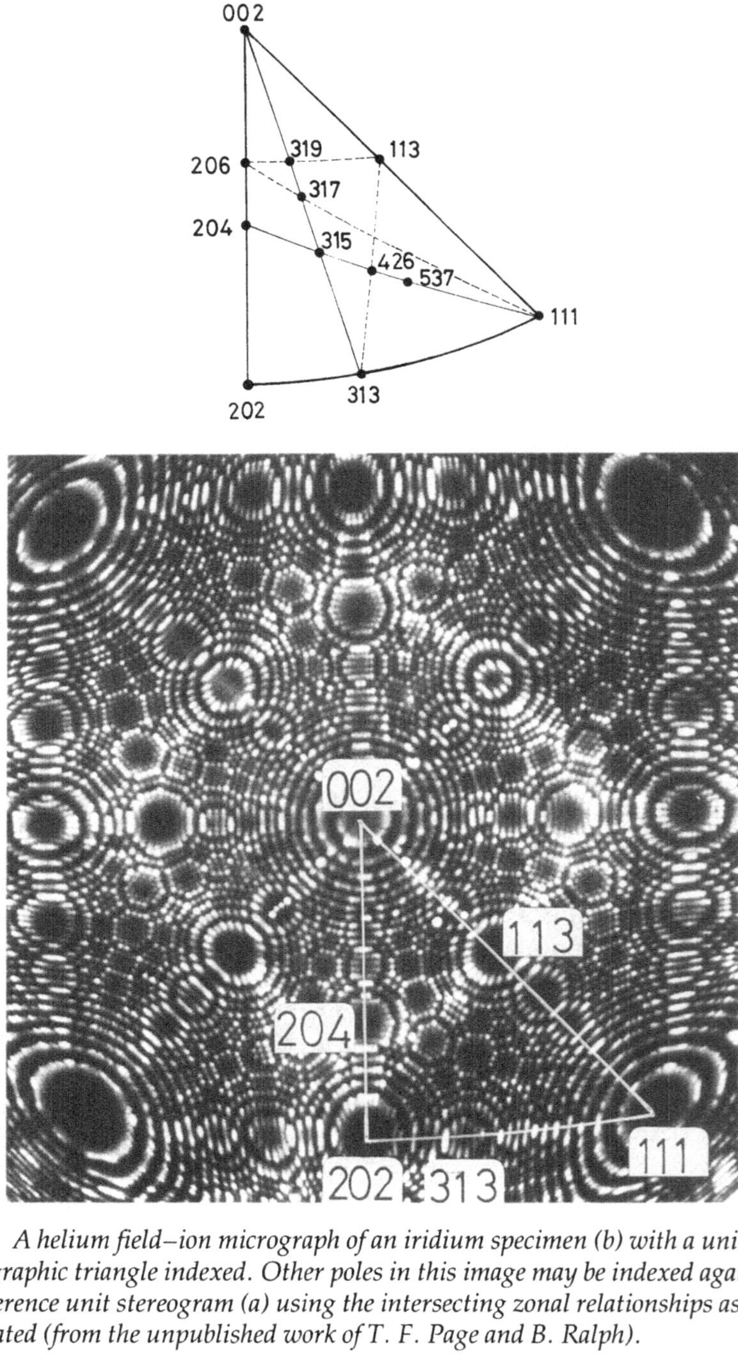

(b)

Fig. 2. A helium field–ion micrograph of an iridium specimen (b) with a unit stereographic triangle indexed. Other poles in this image may be indexed against the reference unit stereogram (a) using the intersecting zonal relationships as illustrated (from the unpublished work of T. F. Page and B. Ralph).

phase contrast, direct imaging of atomic structures within high resolution electron microscopes has become possible so that now the study of the movement of individual atoms/groups of atoms has become relatively routine. These developments have relied very heavily on the skills of instrument developers together with an enormous sophistication of computer techniques in processing images and in their computer interpretation (e.g. ref. 5).

Much development of defect behaviour and mechanisms was performed by electron microscopy in advance of direct atomic imaging and the use of diffraction contrast techniques is still a vital tool in elucidating the behaviour of groups of defects and of the mechanisms of phase transformations.

The first technique to offer atomic resolution was the field–ion microscope (e.g. ref. 6). Although the technique was initially limited to studies of refractory metals, it was possible to demonstrate the structures of a number of defect types relatively early in its development. The Cambridge group pioneered the application of this technique to the study of defects and mechanisms. The projection nature of the image in a field–ion microscope naturally gives magnified images which contain relevant crystallographic information (Fig. 2); paralleling that available from electron microscopes using electron diffraction. To take one example of this, our current understanding and theoretical development of grain boundary structure relies heavily on the early observations by field–ion microscopy[7] and subsequently on very detailed studies made using electron microscopy (e.g. ref. 8).

The importance of having parallel chemical information has been mentioned above and the advent of the atom-probe field–ion microscope and its imaging variant gave an opportunity to collect information regarding the distribution of atomic species with atomic resolution (still unmatched by other techniques). These techniques (that is the atom-probe and the image-probe) have found significant applications in materials science and technology. Again many of these applications were initially pioneered in the Cambridge group (refs. 9, 10).

One new surface science technique which has found application from biology to engineering materials and has the potentiality of direct atomic resolution is scanning tunnelling microscopy. In many ways, this technique combines aspects of field emission and electron microscopy (e.g. refs. 11, 12).

3. Defect Studies

Modern microscopy plays a key rôle in the characterisation of all classes of defects in engineering solids although these techniques are probably still at their best in dealing with defects in crystalline solids. Much interest now centres on the amorphous and microcrystalline states where image interpretation is still in a developing phase.

3.1 Studies of point defects

The driving force for such studies arose to a large extent from developments required by the nuclear power industry, and Alan Cottrell's transfer from the Harwell laboratories to Cambridge established a need for microscopic studies of point defect populations. Some success was achieved in detecting individual point defects and arrays of them by field–ion microscopy (ref. 13). However, there remains an uncertainty here in interpreting these results since the imaging and microdissection of field–ion specimens is accompanied by extremely high mechanical stresses and the production of artefact surface vacancies[14] (see Fig. 3). Clusters of defects may be imaged by refined techniques in the transmission electron microscope and an example is given in Fig. 4.[15]

3.2 Studies in line defects

The establishment of diffraction contrast theory for electron microscope

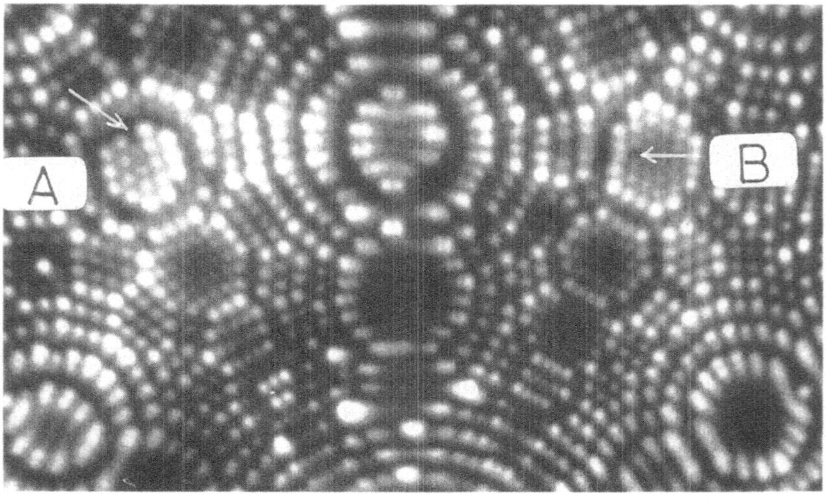

Fig. 3. Examples of surface vacancies on an iridium field–ion specimen created by the high electric field used to image and field evaporate the specimen.[14]

(a)

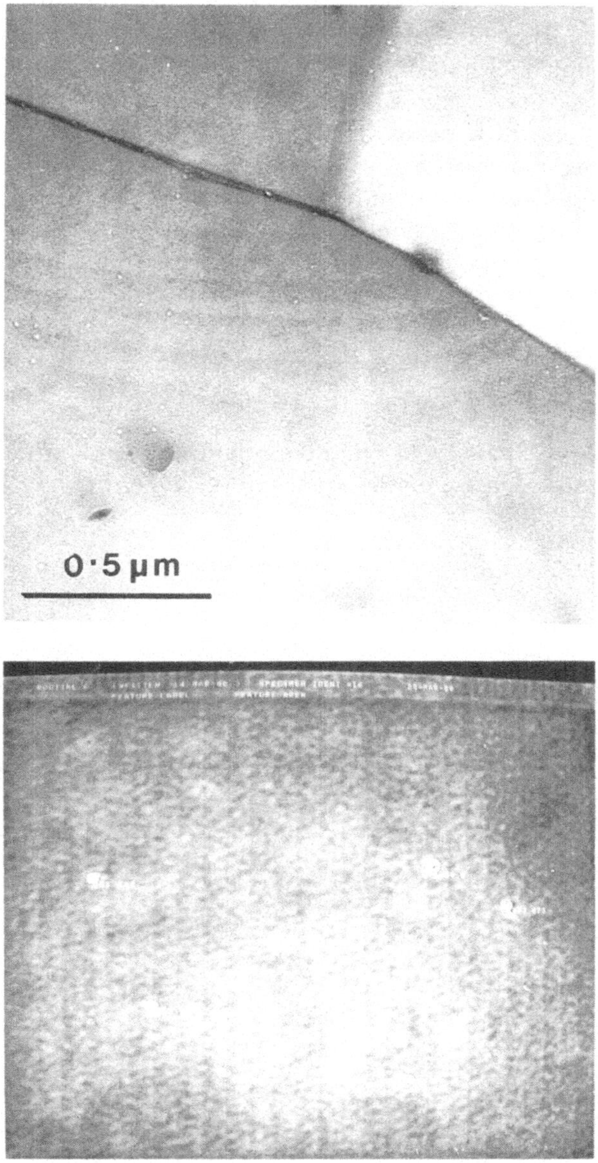

(b)

Fig. 4. (a) Transmission electron micrograph of a Nimonic PE16 sample, after irradiation and annealing to produce a fine array of gas filled bubbles. These bubbles (around 5 nm in diameter) have been revealed by using under-focus to give Fresnel contrast; (b) a similar micrograph but seen after detection on an automatic image analyser.[15]

images allowed the character of dislocations to be determined (e.g. ref. 16). In more recent times, computer processing of direct images of dislocations coupled with the simulation of images, again by computer, has put this crucial area of the subject on a very firm footing (e.g. ref. 5).

The basis on which images from dislocations in field–ion micrographs are interpreted arises from a consideration due to Cottrell.[17] He pointed out that a perfect dislocation intersecting a plane stack at a general angle would turn the stack into a helix. In turn, this is seen as a spiral development of an imaged pole in the field–ion microscope. This concept has been developed to include the contrast from extended dislocations (see Fig. 5).[18]

3.3 *Studies of planar defects*

The original observations of grain boundaries by field–ion microscopy[7] had much to do with the subsequent development of grain boundary

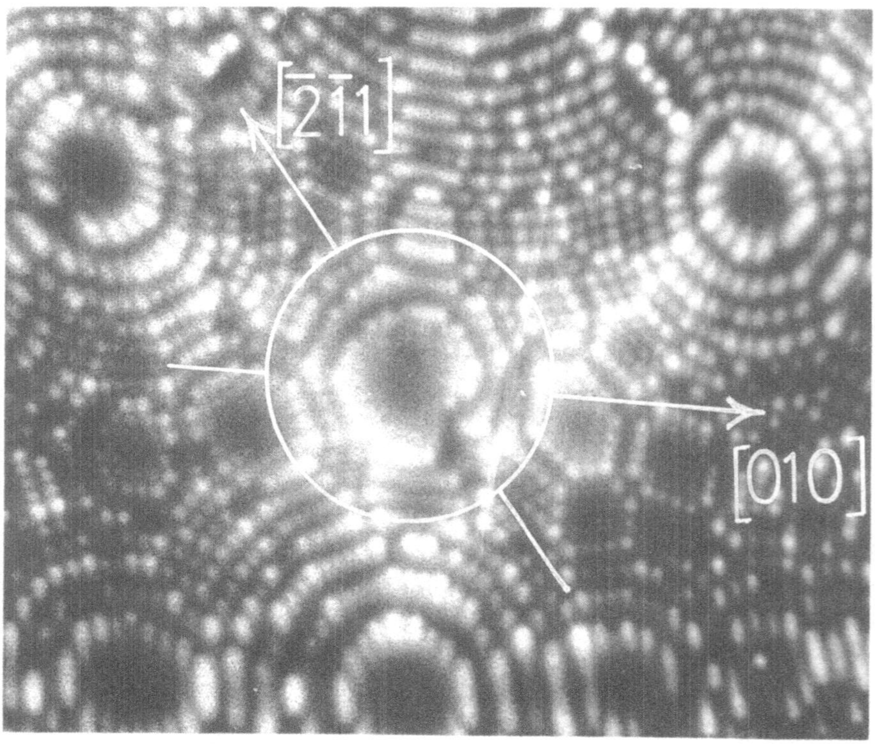

Fig. 5. An extended dislocation in an image from an iridium specimen.[18]

(a)

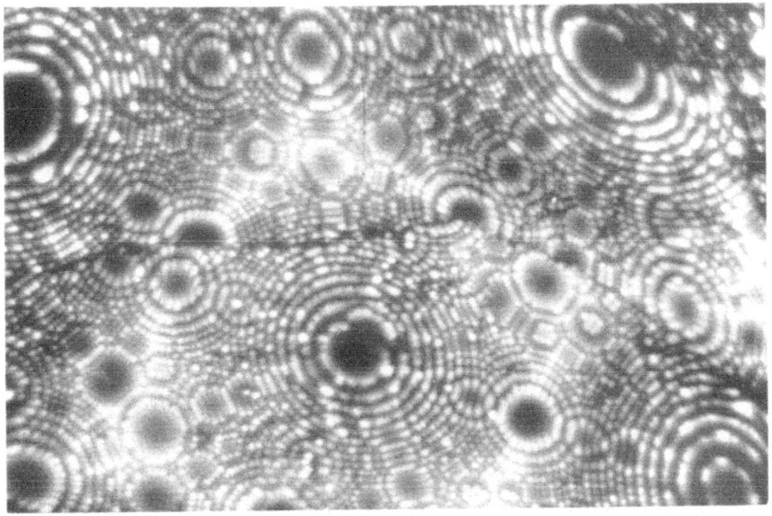

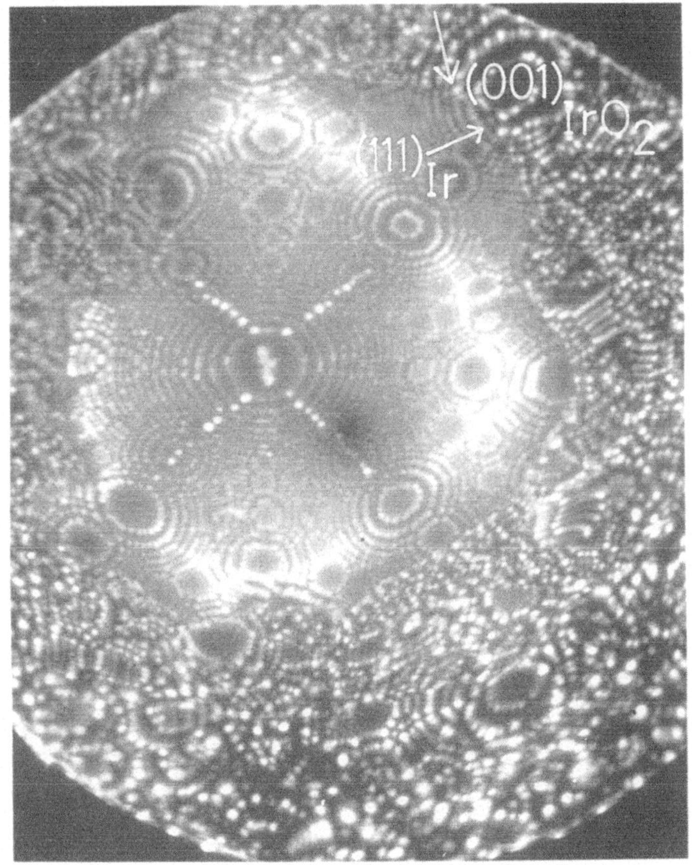

(b)

Fig. 6. (a) A grain boundary in an iridium specimen imaged in the field–ion microscope using helium ions (courtesy T. F. Page); (b) Field–ion micrograph of an oxidized iridium specimen which has been field evaporated to reveal the interface between the metal and the overlying film of polycrystalline IrO_2 *(courtesy C. C. Schubert).*

theory (ref. 19). Figure 6(a) is an example of a field–ion image from a grain-boundary region showing atomic fit right up to the boundary plane. Observations of this type, and the theory which arose from them, lead to the extensive application of transmission electron microscopy in studies of grain boundaries and processes sensitive to and controlled by the presence of grain boundaries (e.g. refs. 8, 20–22).

Far fewer studies have been made by field–ion techniques of interphase interfaces. However, such studies proved valuable in helping to resolve mechanisms of phase transformations, particularly when the data originating from the instrumentation based on field–ion microscopy has been used in a complementary way to that originating from electron microscopy. The surface basis of the field–ion technique makes it particularly valuable for studies of processes at surfaces (e.g. diffusion, corrosion and catalysis). In addition, the nature of thin films has proved to be a particularly fruitful area of investigation and as an example, Figure 6b arises from a study of an IrO_2 oxide film and its interface to the oxidised metal.[23]

4. Mechanistic Investigations

Whilst the studies of defects briefly referred to in the previous main section and subsections are of obvious importance, much microscopical effect is centred upon an understanding of the mechanisms which lead to particular microstructures and substructures. These studies are synergistic with those described before and a few are now presented by way of illustration.

4.1 Studies of phase transformations

A brief look at the proceedings of recent conferences on phase transformations will illustrate the importance of microscopical investigations in elucidating mechanisms (e.g. refs. 24, 25). The examples cited here can only, at the very best, give a flavour of the work undertaken.

Figure 7 arises from a detailed study of the mechanism of cellular decomposition in a copper–titanium alloy.[26] The study here, using transmission electron microscopy, concentrated on the rôle of grain boundaries in the nucleation and growth of the cellular product. Figure 7b is one of a large collection of microdiffraction patterns taken in order to establish the structure of the β phase in this transformation.[27] Microdiffraction in an electron microscope is a relatively new technique for establishing the three-dimensional crystal structure of finely divided phases, the local orientation of regions and may also be used to determine aspects con-

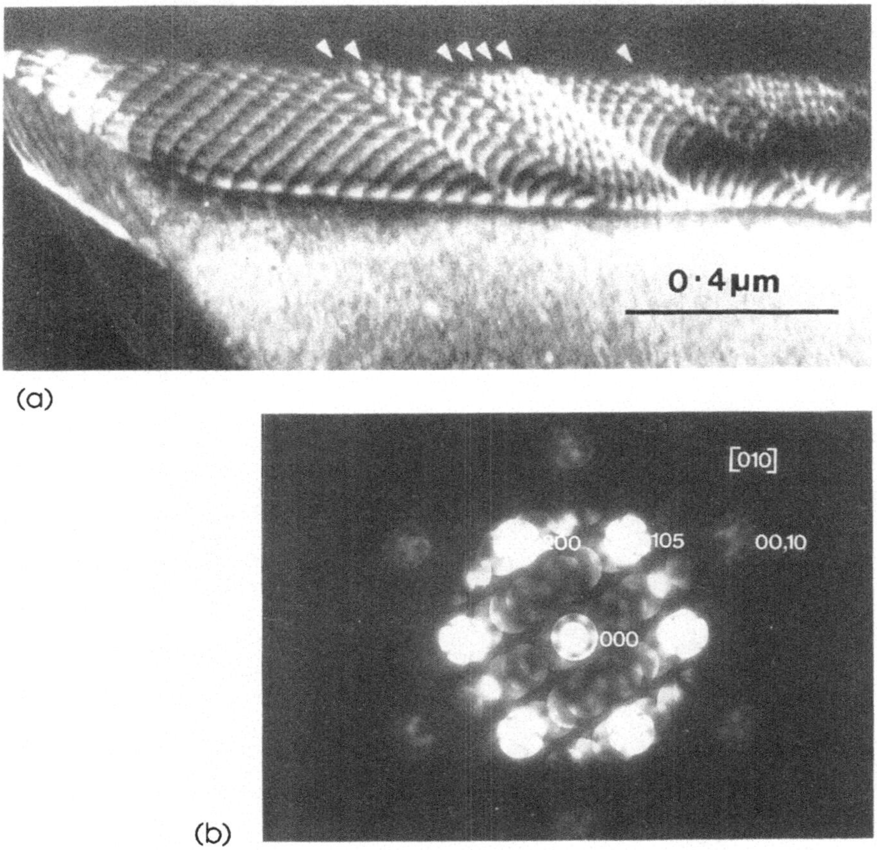

Fig. 7. (a) Weak-beam dark-field image of an α/β interface in a cellularly decomposed copper titanium alloy revealing interfacial dislocations; (b) Microdiffraction pattern from the β phase in the above, used to elucidate the structure.[26,27]

cerned with lcoal chemistry and strain (e.g. ref. 28). Figure 8 arises from a study made of a cellularly formed γ′ precipitate in a nickel-base superalloy which helped to resolve some of the subtle interplays between ordering and the cellular reaction.[29,30]

The final examples considered in this short subsection concern the applications of the atom-probe and imaging-atom probe to studies of phase transformations (e.g. refs. 9, 31). Figure 9 arises from a study of a model nickel-base superalloy (Ni 16·7 at.% Cr 14·5 at.% Al) where the ternary composition has been adjusted to give a low misfit between γ and γ′. The decomposition of the solid solution has been studied after ageing

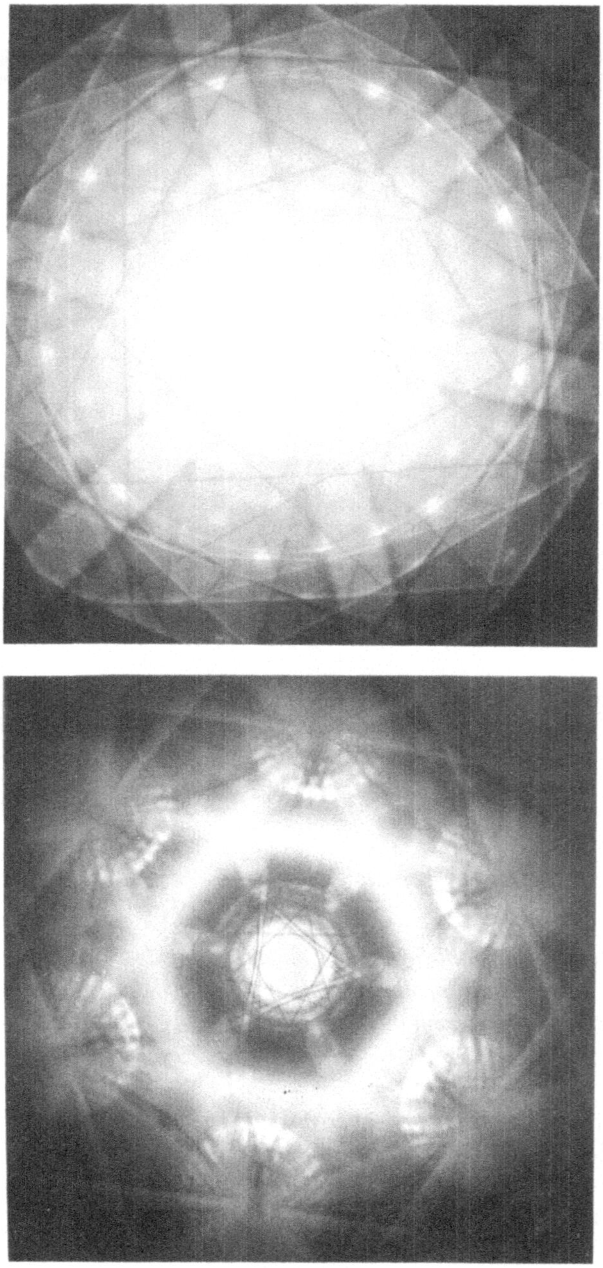

(a)

(b)

Fig. 8. Microdiffraction patterns from a γ' precipitate in a sample of Nimonic alloy 115 in the commercially heat treated condition: (a) gives the whole pattern whilst (b) gives a part of the zero order Laue section and shows the interaction of the upper Laue zone diffraction events in the zero order disc.[28–30]

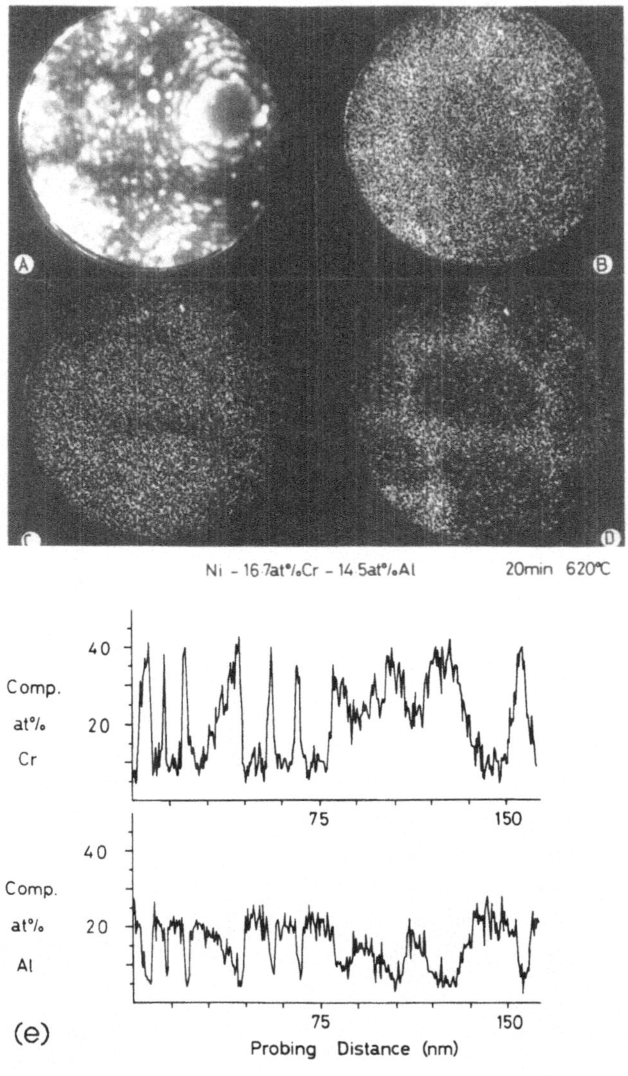

Fig. 9. Field–ion microscope, atom-probe field–ion microscope and imaging atom-probe data from samples of a model ternay alloy (Ni–16·7 atom % Cr–14·5% Al) aged at 620 °C (for 5 min in Figs. 9(a–d) and for 20 min in Fig. 9(e)): (a) The field–ion image (taken in an imaging probe) shows variations in contrast, some of which may be related to composition excursions seen in the gated images (b)–(d) (the approximate field-of-view in images (a)–(d) is 30 nm); (b) is gated for Ni^{24}, *(c) for* Al^{2+} *and (d) for* Cr^{2+}; *(e) gives composition* vs *distance plots taken using an atom-probe.*

(Courtesy S. A. Hill and A. R. Waugh.)[9, 31]

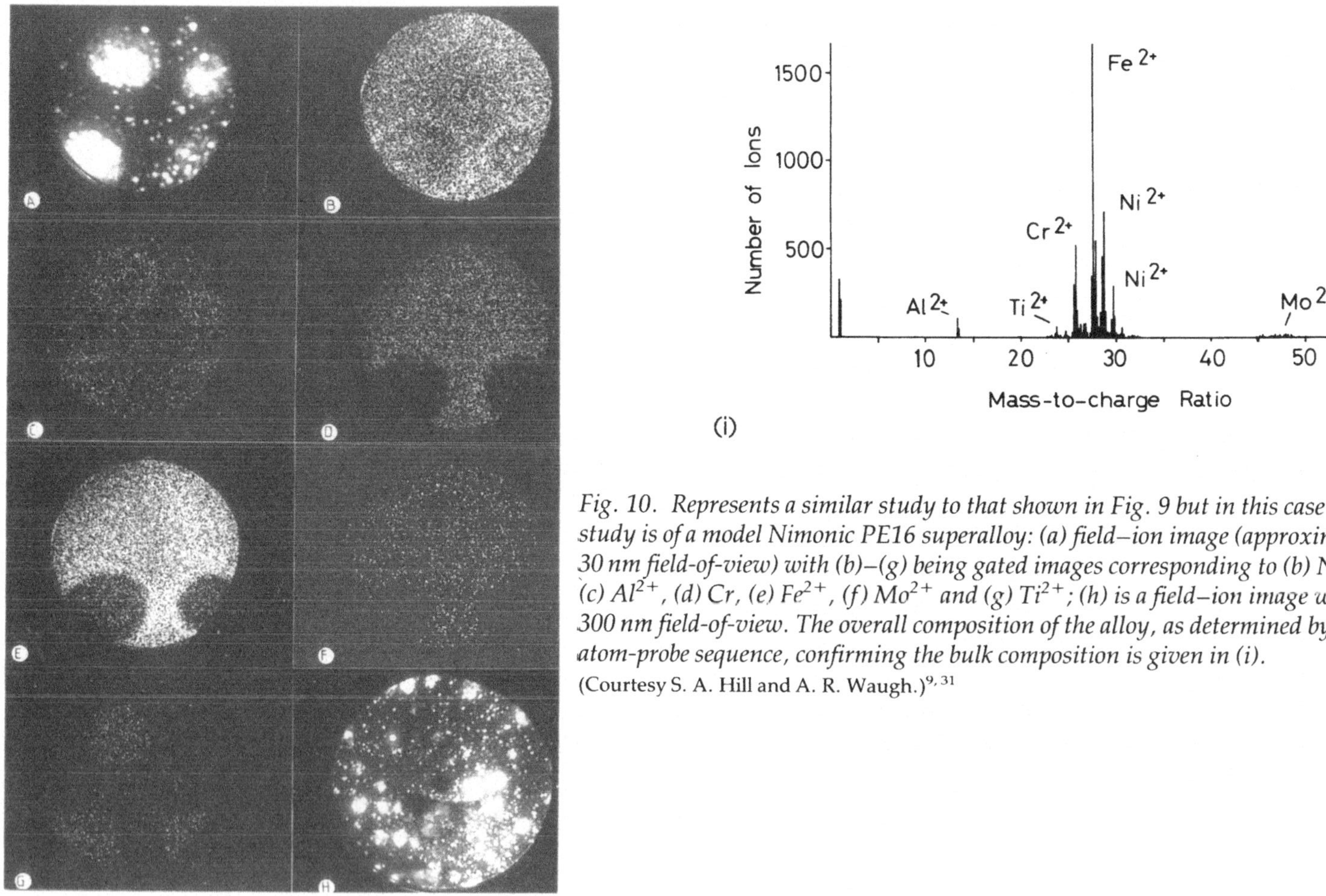

Fig. 10. Represents a similar study to that shown in Fig. 9 but in this case the study is of a model Nimonic PE16 superalloy: (a) field–ion image (approximately 30 nm field-of-view) with (b)–(g) being gated images corresponding to (b) Ni^{2+}*; (c)* Al^{2+}*, (d) Cr, (e)* Fe^{2+}*, (f)* Mo^{2+} *and (g)* Ti^{2+}*; (h) is a field–ion image with a 300 nm field-of-view. The overall composition of the alloy, as determined by an atom-probe sequence, confirming the bulk composition is given in (i).*
(Courtesy S. A. Hill and A. R. Waugh.)[9, 31]

at 620 °C. Figure 9(a) is a conventional field–ion micrograph of a 30 nm field of view of a specimen aged for 5 minutes. Enhanced contrast is seen from regions which are tending towards the γ' composition. Figures 9(b–d) are images gated for particular atomic species (see caption) and these illustrate the overall partitioning. By contrast, Figure 9e, presents atom-probe composition profiles as a function of distance for a specimen aged for 20 min. It is apparent that the aluminium and chromium excursions tend to go in antiphase. The outcome of the study suggested a continuous transformation mechanism.

By contrast, similar studies of the decomposition of an alloy based on the Nimonic PE16 composition demonstrated that the production of γ' here was by a nucleation and growth mechanism.[9,31] Figure 10 arises from this study with Figure 10(a–g) having a field-of-view of around 30 nm and 10(h) around 300 nm. Figures 10(a) and (h) are conventional field–ion images taken in an imaging-probe and atom-probe field–ion microscopes respectively. The distribution of atomic species is shown in the gated images (Figs. 10(b)–(g) which clearly suggests the presence of stable γ'. Figure 10(i) gives the overall composition of the specimen as determined by atom-probe field–ion microscopy, and it matches well with the nominal composition of the bulk alloy. This particular specimen had received a complex heat treatment; solution treated at 1,050 °C for 2 h, aged at 900 °C for 1 h then 750 °C for 8 h to simulate commercial practice.

4.2 *Studies of deformation substructures*

An understanding of the process of deformation by slip comes not only from the properties of single dislocations but also from how groups of dislocations behave, how they interact with other microstructural features, such as grain boundaries, and how the deformation substructure builds up with continued strain.[31] Some aspects of the interaction of dislocation interactions with grain boundaries have been studied by field–ion microscopy[32] but the small field-of-view of this technique and the high field-induced mechanical stresses limit such studies. By contrast, transmission electron microscopy has proved to be an admirable technique for such studies and again these observations have been used to underpin the theoretical development of deformation processes.

Figure 11 illustrates recent observations by transmission electron microscopy of deformation substructures. Figure 11(a) illustrates a planar pile up of dislocation interacting with the intrinsic grain boundary dislocation structure in a near $\Sigma = 9$ coincidence grain boundary in a nickel base superalloy sample. Our understanding of the details of the Hall–Petch relationship may well come from observations of this type. As deformation proceeds in materials with higher staking fault energies, the

(a) 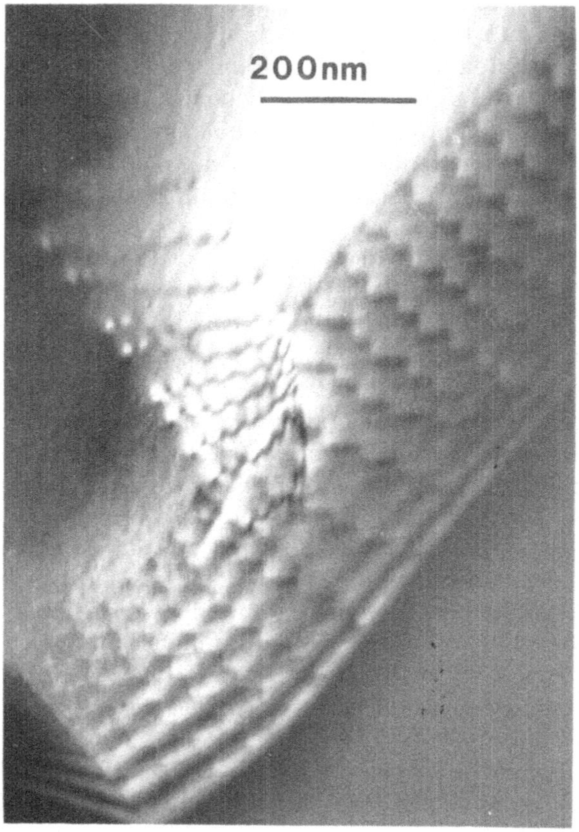

Fig. 11. (a) Transmission electron micrograph of a dislocation pile up entering a grain boundary in Nimonic PE16 and interacting with the grain boundary dislocation structure in a near $\Sigma = 9$ boundary;[21] *(b) A montage of transmission electron micrographs from a region in a pure aluminium sample containing a grain edge and after straining to $\epsilon = 0{\cdot}10$. (b(II)) gives the misorientations in degrees across the cell walls imaged in (b(I)), as determined by microdiffraction.*[33]

deformation pattern builds into a cell structure[33] and may eventually incorporate various forms of bands.[34] Figure 11(b(i)) illustrates the cell structures developed near a grain boundary node in a pure aluminium sample strained to $\epsilon = 0{\cdot}1$. Microdiffraction has been used to establish the local orientation of the cells and Figure 11(b(ii)) gives the angular misorientations in degrees, across the cell walls.

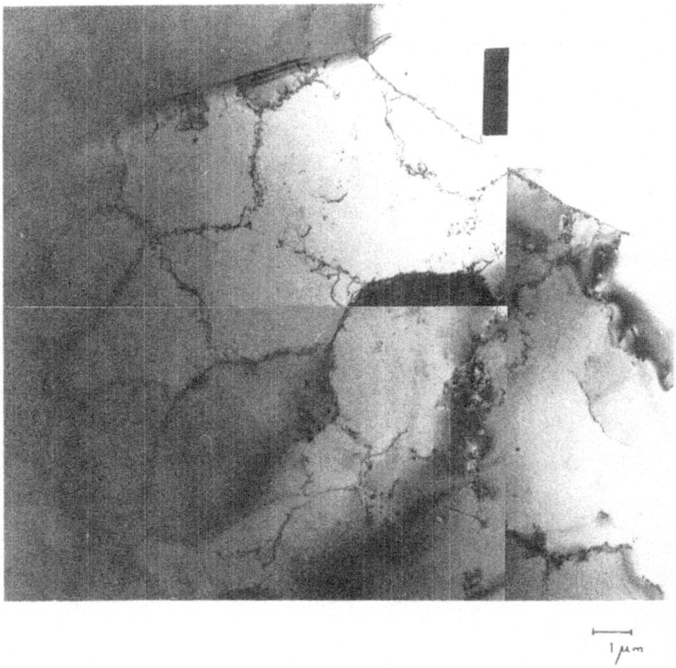

(b (i))

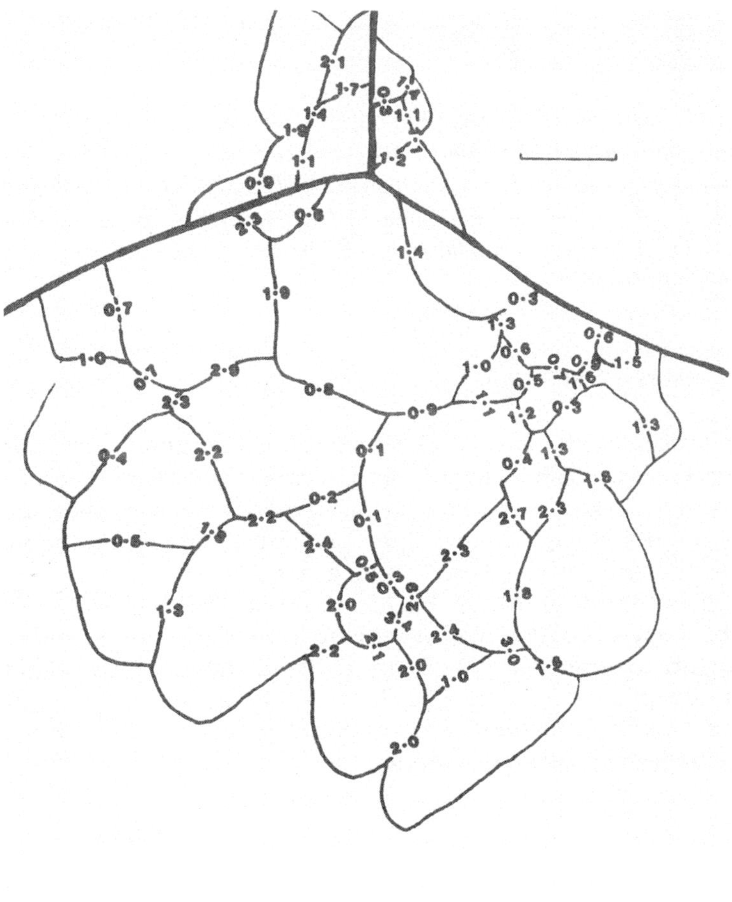

(b (ii))

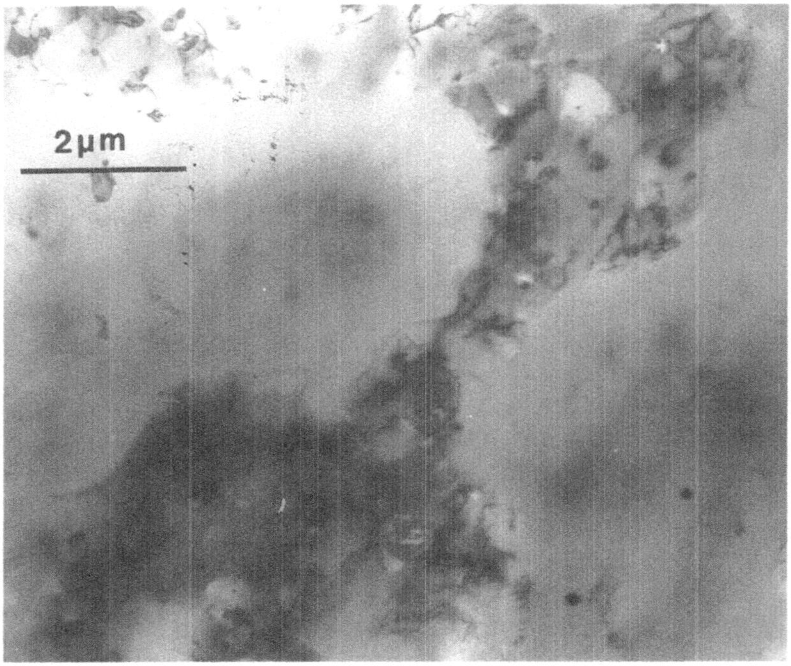

(a)

Fig. 12. (a) Transmission electron micrograph from a commercial 8091 aluminium–lithium alloy sample after cold work and partial recrystallisation (from unpublished work of S. Wismayer and B. Ralph); (b) Transmission electron micrograph from a Nimonic PE16 sample where the γ' population in contact with the grain boundary has coarsened and stopped the grain boundary from migrating.[39]

4.3 *Studies of heat treatment*

The processes of recovery, recrystallisation and grain growth are of key importance in establishing the optimum microstructure in an alloy to give a chosen basket of mechanical properties. Recent studies, mainly by electron microscopy, have helped to reinforce and refine the theoretical development of these processes (e.g. refs. 35–40).

Much recent effort has been centred on the mechanisms by which viable recrystallisation nucleii develop from the deformed and recovered structure (e.g. refs. 36, 37). Figure 12(a) illustrates one aspect of this type of study for a modern airframe alloy (8091 based on aluminium–lithium). Here copious nucleation of recrystallisation has occurred by a variant of the sub-grain coalescence model.

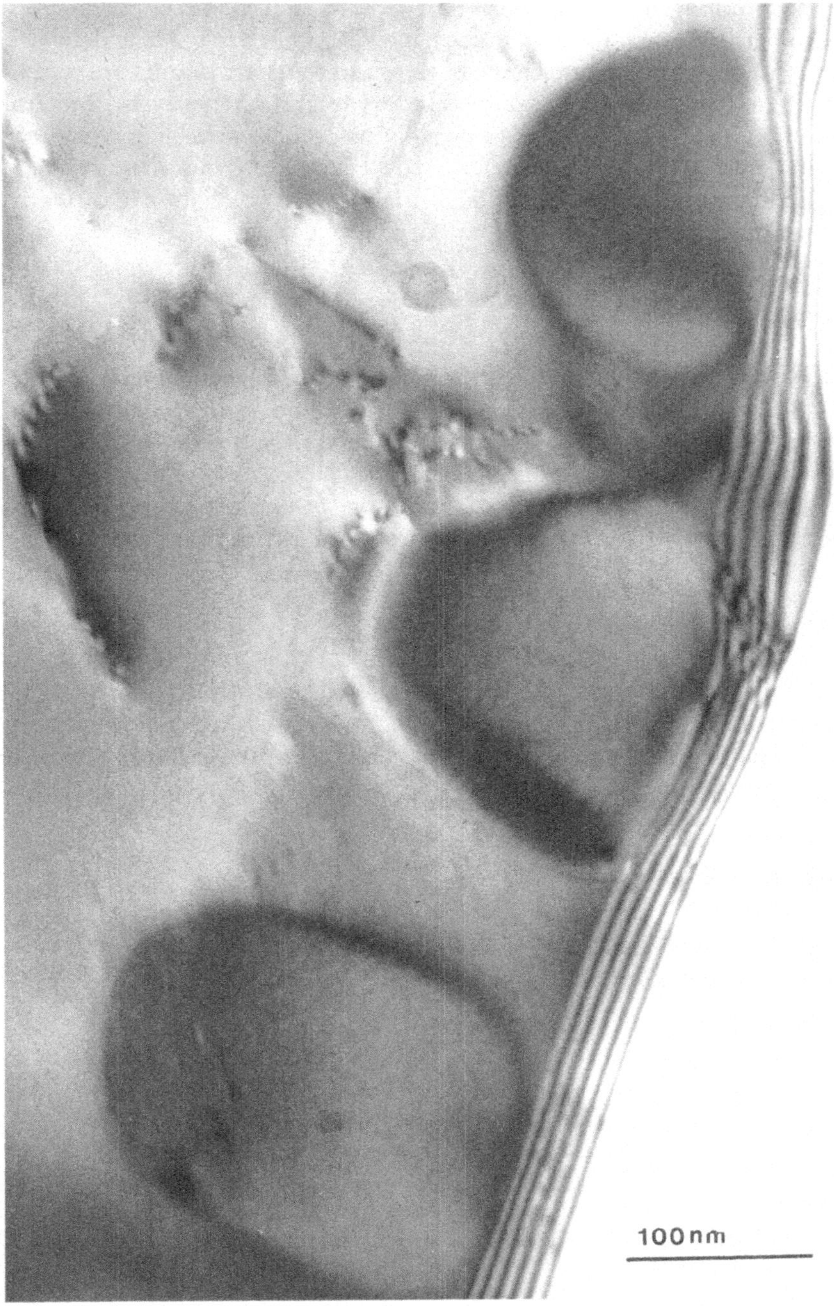

(b)

Controlling the process of grain growth is of vital technological importance and many recent studies have examined this in some detail (ref. 40). Coherent precipitates have been found to have a significant pinning effect, to the point where grain boundary migration may cease and grain boundary sliding, to produce a population of more-special, lower energy grain boundaries may occur[22, 39] (see Fig. 12(b). This may present a method of producing polycrystals with further modification to properties *via* a route — termed grain boundary engineering.[8]

5. Concluding Remarks

Much of the driving force for the type of microstructural investigation cited here, albeit very briefly, arises from a quest to understand how microstructure evolves and how this then controls properties. Alan Cottrell, as an outstanding example of a scientist involved with the theoretical development of this subject, should be seen as of key importance in providing the reason for many of these microscopical investigations.

Acknowledgements

It is a pleasure to record my indebtedness to Alan Cottrell for what has been a long and inspiring association. I also wish to acknowledge my appreciation of many colleagues who have contributed to this brief review.

References

1. A. H. Cottrell, *J. Inst. Met.* **90**, 449–456, 1962.
2. P. G. Careres, B. Ralph, G. C. Allen and R. K. Wild, *Phil. Mag.* **59**, 1119–1162, 1989.
3. B. Ralph, *Proc. Roy. Inst.* **61**, 251–281, 1989 (in press).
4. G. A. Bassett, D. W. Pashley and J. W. Menter, *Proc. Roy. Soc.* **A246**, 345–368, 1958.
5. M. H. Loretto, in Conf. Dislocations and Properties of Real Materials (Inst. of Met. No. 323, London), pp. 15–26, 1984.
6. E. W. Müller and T. T. Tsong, Field–Ion Microscopy — Principles and Applications. Elsevier, New York, 1969.
7. D. G. Brandon, B. Ralph, S. Ranganathan and M. S. Wald, *Acta Met.* **12**, 813–821, 1964.
8. B. Ralph, *Univ. of Wales Review*, **4**, 29–38, 1988.
9. B. Ralph, S. A. Hill, M. J. Southon, M. P. Thomas and A. R. Waugh, *Ultramicroscopy*, **8**, 361–375, 1982.

10. A. Cerezo, T. J. Godfrey, C. R. M. Grovenor, M. G. Hetherington, R. W. Hoyle, J. P. Jakubovics, J. A. Liddle, G. D. W. Smith and G. M. Worrall, *J. Microsc.* **154**, 215–225, 1989.
11. B. Ralph, *J. Microsc.* **155**, 339–360, 1989.
12. P. K. Hansma and J. Tersoff, *J. Appl. Phys.* **61**, R1–R23, 1987.
13. D. N. Seidman, *Surf. Sci.* **70**, 532–565, 1978.
14. T. F. Page and B. Ralph, *Surf. Sci.* **36**, 9–28, 1973.
15. M. P. Shaw, V. L. Kohler and B. Ralph. *J. Microsc.* **128**, 251–259, 1982.
16. P. B. Hirsch, A. Howie, R. B. Nicholson, D. W. Pashley and M. J. Wheland, *Electron Microscopy of Thin Crystals*, Butterworths, London, 1965.
17. A. H. Cottrell, *Theory of Crystal Dislocations*. Blackie, London, 1964.
18. D. A. Smith, T. F. Page and B. Ralph. *Phil. Mag.* **19**, 231–240, 1969.
19. W. Bollman, *Crystal Defects and Crystalline Interfaces*. Springer-Verlag, Berlin, 1970.
20. B. Ralph, *Journal de Physique*, **C4**, 71–79, 1975.
21. V. Randle and B. Ralph, *Revue Phys. Appl.* **23**, 501–512, 1988.
22. V. Randle and B. Ralph, *Proc. Roy. Soc.* **A415**, 239–256, 1988.
23. M. A. Fortes and B. Ralph, *Proc. Roy. Soc.* **A307**, 431–448, 1968.
24. H. I. Aaronson, D. E. Laughlin and R. F. Sekerka (Eds), Inst. Conf. Solid → Solid Phase Transformations, Met. Soc. AIME, Warrendale, PA, 1982.
25. G. W. Lorimer (Ed.) Int. Conf. Phase Transformations, Inst. of Metals, No. 404, London, 1988.
26. R. C. Ecob, J. V. Bee and B. Ralph, *Met. Trans.* **11A**, 1407–1414, 1980.
27. R. C. Ecob, J. V. Bee and B. Ralph, *J. Microsc.* **119**, 153–161, 1980.
28. B. Ralph and R. C. Ecob, in Proc. 5th Int. Risø Symp., Techniques, 109–129, 1984.
29. R. C. Ecob, M. P. Shaw, A. J. Porter and B. Ralph, *Phil. Mag.* **44**, 1117–1133, 1981.
30. A. J. Porter, M. P. Shaw, R. C. Ecob and B. Ralph, *Phil. Mag.* **44**, 1135–1148, 1981.
31. B. Ralph, *Proc. Roy. Mic. Soc.* **15**, 254–266, 1980.
32. T. F. Page and B. Ralph, *Surface Sci.* **23**, 211–224, 1970.
33. I. Barker, N. Hansen and B. Ralph, *Mat. Sci. & Eng.* **A113**, 449–454, 1989.
34. B. Bay and N. Hansen, *Mat. Sci. & Eng.* **A113**, 385–397, 1989.
35. A. R. Jones, B. Ralph and N. Hansen, *Met. Sci. J.* **13**, 149–154, 1979.
36. A. R. Jones, B. Ralph and N. Hansen, *Proc. Roy. Soc.* **A368**, 345–357, 1979.
37. B. Ralph, C. Y. Barlow, B. A. Cooke and A. J. Porter, Proc. 1st Int. Risø Symp. Recrystallisation, 229–242, 1980.
38. C. J. Tweed, N. Hansen and B. Ralph, *Met. Trans.* **14A**, 2235–2245, 1983.
39. V. Randle and B. Ralph, *Acta Met.* **34**, 891–898, 1986.
40. V. Randle, B. Ralph and N. Hansen, Proc. 6th Int. Risø Symp., Annealing, 123–142, 1986.

9 Creep and Creep Fracture

G. W. Greenwood

Professor, Division of Metals School of Materials, University of Sheffield

Contribution to the Scientific Meeting in honour of Sir Alan Cottrell, 29 September 1989

Abstract

The review focuses upon the interpretation of experimental work and understanding of the mechanisms of creep and creep fracture that relate to Alan Cottrell's early studies and perceptions.

Attention is first given to the time laws of creep and to diversity in the form of transients that follow sudden stress changes. Here concepts of solute drag on dislocation motion and of the tendency to balance hardening and recovery processes remain important and can be extended to account for a variety of observations. In steady state situations the creep rate variation with stress can often be understood in terms of different creep mechanisms when account is taken of their interaction and transition regions.

Next the role of internally generated stresses is considered in causing pseudo creep under small external stress. This feature has found increasingly wide application in relation to the thermal cycling of composite or polycrystalline materials comprising crystals with anisotropic properties. A similar approach can also account for deformation during phase transformations. Superplastic behaviour has sometimes resulted from the linear dependence of pseudo creep on applied stress.

Finally the occurrence of creep fracture after small strains is examined. In such cases it is necessary to distinguish mechanisms of crack formation and growth through stress concentrations from those of cavity growth through vacancy diffusion and condensation with plastic flow sometimes playing an additional role. The specific conditions for cavity nucleation remain difficult to resolve. Models of crack and cavity growth are becoming well developed and much of practical value is being gained through an understanding of ways in which different mechanisms can be linked in series or in parallel operation.

Introduction

The distinctive and significant contributions that Alan Cottrell made to our understanding of creep and creep fracture may conveniently be placed into 3 groups:

(1) The time laws of creep and effects of stress changes.
(2) The influence of internally generated stresses on creep.
(3) The processes leading to creep fracture.

The contributions are all characterised by his viewing of these areas within a broader context, sometimes identifying and developing key features of studies previously ignored or insufficiently recognised and sometimes laying new foundations where these did not exist or were insecure. The features on which he laid emphasis have led to extensive further developments, frequently revealing a widespread usefulness that would have been difficult to foresee.

An attempt is made here to appraise some of these developments in relation to the earlier work and to show that, despite many remaining uncertainties, clearer pictures are emerging of underlying mechanisms. Their value is becoming increasingly recognised in practical situations where significant features and predictions are more realistically and confidently being taken into account.

The Time Laws of Creep

Strain-time relationships

It has long been noted[1,2] that, after some instantaneous strain ϵ_0 on loading, the variation of creep strain ϵ with time t may display three (primary, secondary and tertiary) regions. In the primary stage, Andrade[3] first found $\epsilon \alpha t^{1/3}$ and this result has since been obtained[4] for a remarkable diversity of materials, including metal single crystals[5] and some polymers,[6] suggesting it must be interpreted in quite general terms. Creep strain in the secondary stage, by definition, increases linearly with time, whilst in the tertiary stage the creep rate accelerates. Many expressions have been proposed[7] to describe the complete creep curve at constant stress and temperature and quite a good fit has often been found[8] in an equation of the form:

$$\epsilon = \epsilon_0 + \beta t^{1/3} + Kt + \gamma t^3 \tag{1}$$

where β, K and γ are constants. Nevertheless the search for improved formulation has proceeded relentlessly over several decades. The $t^{1/3}$ relationship remains without a satisfactory explanation unless special features of work hardening are assumed.[9] However, wide ranging studies greatly extending experimental data to lower homologous temperatures and longer creep lives, have tended to suggest a logarithmic expression is then more widely applicable with $\epsilon \alpha \ell n(1 + ct)$, where c is a constant, which is readily reconcilable with theories of hardening and recovery. Putting this in a nearly equivalent exponential form together with more detailed analysis of tertiary creep data has led to further proposals such as[10]

$$\epsilon = \epsilon_o + \epsilon_1[1 - \exp(-t/\tau_1)] + Kt + \epsilon_3\exp[t - t_f)/\tau_3] \qquad (2)$$

where ϵ_1 and ϵ_3 represent primary and tertiary creep strains and t_f is the time to failure. τ_1 and τ_2 are effectively time constants for primary and tertiary creep. Bearing in mind the variation of all these parameters with stress and temperature, the massive data collection required to form an empirical basis for creep life prediction soon becomes apparent.

Bold attempts have been made to minimise this problem,[11] principally through the suggestion that the secondary creep stage is the least significant from the point of view of formulation, in so far as the addition of merely two terms covering primary and tertiary creep may give a sufficiently long region where, approximately, the creep rate can be considered constant. This approach[12] has led to the suggestion that:

$$\epsilon = \epsilon_0 + \theta_1[1 - \exp(-\theta_2 t)] + \theta_3[\exp(\theta_4 t) - 1] \qquad (3)$$

where each of the four θ parameters characterises the material but with each relying on a further four parameters to account for their stress and temperature variation. Changes in chemical composition, thermal history, environmental conditions, microstructure, possible anisotropy and the effects of stress and temperature transients all demand still further information[13] and it is against this background that Cottrell's early search for some improved understanding must be viewed. In particular he recognised the importance of previous indications that behaviour following stress transients was likely to be one of the key features in obtaining a clearer insight into the mechanisms of creep.

Transients in creep

Cottrell's studies, some 40 years ago, of the flow of zinc single crystals under constant shear stress and of polycrystals under constant tensile stress at elevated temperature[5] proved generally in accord with Bailey and Orowan's earlier views[14, 15] that creep rate can be expected to be related to a coefficient of strain hardening and to the rate of recovery.

Especially supportive of such a view was the experimental observation that, after sudden stress reduction, there would be a period during which the creep rate was small or even zero before increasing to a new steady state characterised by the new stress level.

Such an approach has since been extensively explored,[16–18] resulting in the experimental identification of transient periods where strain–time relationships have taken various forms dependent both on the material and on the creep conditions. It is only recently that it has appeared possible to interpret such diversity within a more general framework which offers promise of incorporating all the principal features of transient effects. The analysis,[19] summarised below, is based on earlier models[20] with their progressive extension of the concept of recovery creep where the dislocation substructure is governed by the tendency to balance the rate of dislocation creation with the rate of their annihilation and with a dynamic equilibrium established during the secondary creep stage.

It may be considered that this balance is temporarily destroyed by a sudden stress change. The nature of the dislocation network prior to and following the change is responsible for the variety in the form of the creep curves reported[7] during the transient period, which represents the time required to establish the new equilibrium. As well as the situation already mentioned, where an incubation period may follow a stress drop during which the creep rate is zero, in other cases the creep rate may be initially greater or less than that which eventually corresponds to the new stress level.[21] It is also possible to interpret the situation in which the creep rate can become negative through an anelastic contribution.

It is implicit in the newer approaches[19] that features such as the drag produced by dislocation interaction with solute atmospheres, developed[22, 23] from the original analysis[24] of Cottrell and Jaswon remain relevant. It is proposed that a dislocation network is generated in creep, with an average spacing λ between nodes, through dislocations gliding under applied stress σ with a velocity V_g given by

$$V_g = M_g b^2[\sigma - (Gb/\lambda)] \tag{4}$$

where M_g is the dislocation glide mobility, b is the Burgers vector and G is the shear modulus of the material. In equation (4), Gb/λ may be regarded as the average retarding stress due to other dislocations such that glide is not possible under an applied stress below this value. Now $V_g \sim \lambda/t_c$, where t_c is the time to increase the dislocation density ρ to $1/\lambda^2$, so that the rate of dislocation creation can be written:

$$(\mathrm{d}\rho/\mathrm{d}t)_c = 1/\lambda^2 t_c \sim 2V_g/\lambda^3 = (2M_g b^2/\lambda^3)[\sigma - (Gb/\lambda)] \tag{5}$$

It is considered that dislocations are continually annihilated by diffusional processes and, following Friedel,[25] the dislocation network coarsens with a rate of decrease of dislocation density

$$(d\rho/dt)_a = -2DGb^3/\lambda^4 kT \quad (6)$$

at temperature T where D is a diffusion coefficient and k is Boltzmann's constant.

When the secondary stage of creep is reached, a balance is struck between the rates of dislocation creation and annihilation so that $(d\rho/dt)_c + (d\rho/dt)_a = 0$. Hence, from equations (5) and (6) we obtain:

$$M_g[\sigma - (Gb/\lambda)] = DG/\lambda kT \quad (7)$$

Now to satisfy equation (7), λ may be denoted by λ_s corresponding to the steady state value, given explicitly by rearranging that equation in the form:

$$\lambda_s = (Gb/\sigma)[(kT/DM_g) + 1] \quad (8)$$

Here it is noted that D/kT is the dislocation climb mobility M_c and, following Burton,[19] it is convenient to introduce a parameter $\alpha = M_g/M_c$ which is the ratio of the dislocation glide to climb mobilities.

The creep rate at the steady state is given[15] by $\dot{\epsilon}_s = bV_g/\lambda_s^2$ and so substituting for V_g from equation (4) and for λ_s from equation (8) we obtain:

$$\dot{\epsilon}_s kT/DGb = (\sigma/G)^3[\alpha/(1 + \alpha)]^3 \quad (9)$$

It is noted this formulation is comparable with the type of equation proposed by Dorn,[26, 27] which will be referred to later.

The creep rate $\dot{\epsilon}$ can be described for any value of λ directly from equation (4) with appropriate substitutions and rearrangement, leading to the more general situation where

$$\dot{\epsilon} kT/DGb = \alpha(b/\lambda)^2[\sigma - (Gb/\lambda)]/G \quad (10)$$

It is informative to compare the steady state situation represented by equation (9) with the more general case in equation (10), and the latter equation is illustrated graphically in Figs. 1 and 2. The creep rate is zero below some critical value of λ and reaches a maximum $\dot{\epsilon}_{max}$ for the value $\lambda = 3Gb/2\sigma$. Substituting this value of λ in equation (10) shows that

$$\dot{\epsilon}_{max} = (4\alpha/27)(DGb/kT)(\sigma/G)^3 \quad (11)$$

which is only equal to the steady state creep rate $\dot{\epsilon}_s$ when $\alpha = 2$. More generally:

$$\dot{\epsilon}_s/\dot{\epsilon}_{max} = 27\alpha^2/4(1 + \alpha)^3 \quad (12)$$

For solute drag alloys of equal glide and climb mobilities with $\alpha = 1$, we

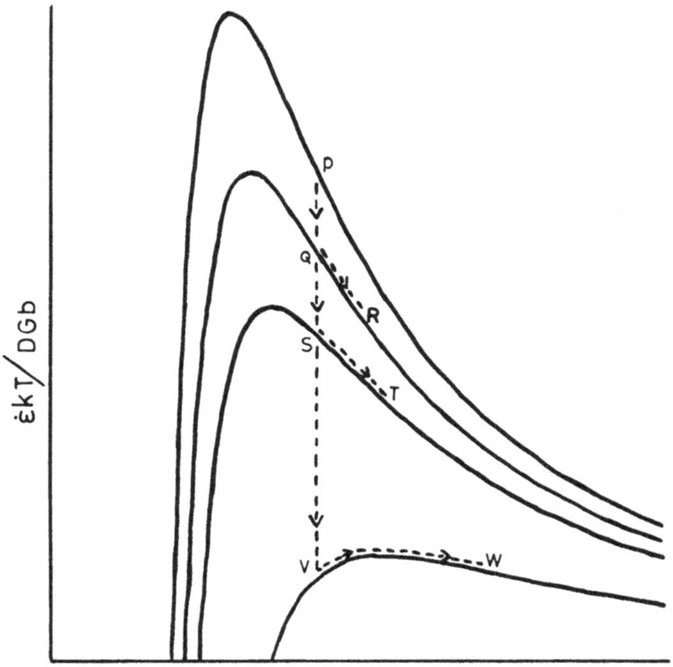

Fig. 1. The variation of creep rate $\dot{\epsilon}$ with dislocation network spacing λ plotted in dimensionless form, according to equation (10). The upper curve is for the highest stress level and from equation (9) the steady state creep rate $\dot{\epsilon}_s$ for a solute drag alloy (class I type behaviour) with $\alpha = 1$ corresponds to the point P to the right of the maximum. On sudden stress reduction the creep rate falls to the value at point Q on the curve relating to the new (lower) stress level, and decreases until point R is reached where the dislocation network λ again corresponds to the steady state. The situation for a stress reduction leading to a creep rate drop from P to S is similar, reaching a steady state at point T. Both these cases produce transients of the form of a new primary creep stage. For a larger stress drop the creep rate may correspond to a point at V. In this case there may be a period of increasing creep rate before the subsequent decrease to the steady state at W.

have a situation depicted in Fig. 1 where the dislocation network spacing λ is relatively large and the creep rate corresponds to the point P to the right of the maximum on the upper curve relating to the highest stress level. If the stress is suddenly reduced to the point Q on the curve relating to this new (lower) stress, the network size than gradually enlarges and the creep rate progressively decreases to reach a new steady state value depicted at *R*. A similar situation appertains on a stress drop to *S* where the new steady state creep rate corresponding to *T* is approached. Both these cases show that with such stress reductions, a new but conventionally shaped primary creep curve results. The situation, however, is markedly different for a larger stress drop to the value at *V* which is now at the left hand side of the maximum on the curve for the new stress level. In this case there is first a gradual increase in creep rate followed by a reduction which stabilises at the steady state value at *W*. The form of transient creep behaviour following a substantial stress drop has been

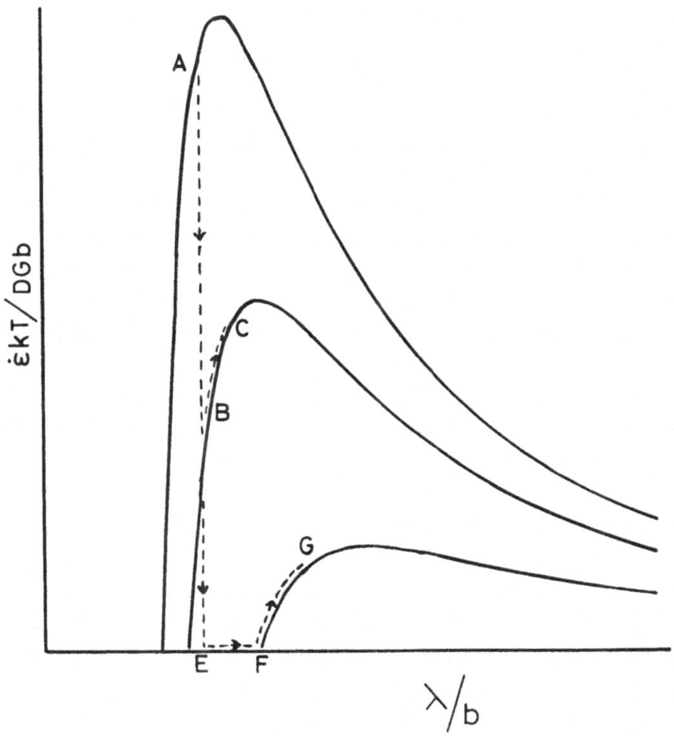

Fig. 2. These curves are again drawn from equation (10). For a rapid glide alloy (class II type behaviour), taking $\alpha = 5$, the steady state creep rate $\dot{\varepsilon}_s$ from equation (9) now corresponds to point A to the left of the maximum. On sudden stress reduction the creep rate is immediately reduced to the value at point B but gradually increases to the value at C corresponding to the new steady state value. If the stress reduction from A is larger, it may give the situation at E where it is initially insufficient to sustain creep. There is then an incubation period until the point F is reached where the dislocation network has coarsened sufficiently for creep to restart. The creep rate subsequently progressively increases until point G is reached.

termed 'inverse primary creep' for it comprises an initial increase in creep rate with a subsequent decrease to give a region on the creep curve of sigmoidal form. This feature is now known[21] to occur in such circumstances for alloys now generally referred to as showing 'Class I' type behaviour.[28] (Care must be taken in referring to such a classification since some authors[29] have adopted a reversal of this classification.) These alloys are characterised by a creep stress exponent of 3 and are usually considered to be influenced by solute drag in creep.

A different form of behaviour is apparent for rapid glide materials, typically pure metals and this is often referred to as showing 'Class II' behaviour[28] where it is appropriate to take a higher value of α, such as $\alpha = 5$. Now it is noted from Fig. 2 that the creep rate corresponds to point A which is to the left hand side of the maximum of the curve relating to the highest stress level. After sudden stress reduction to point B, the lowered creep rate progressively increases until the steady state value is reached at C. With a larger stress dip, the line from A may intersect the abcissa at E and here we have the situation where the creep rate is temporarily zero until the network coarsens sufficiently to become compatible with the new stress level. The period spent in traversing the line E F corresponds to the incubation period before creep begins at F and the rate eventually rises to the steady state value corresponding to G. Such incubation periods have frequently been observed as one of the characteristics of 'Class II' type behaviour[7] and again, as in the previous areas discussed, the envisaged interpretation of many creep phenomena through models of dislocation network recovery creep seems justified.

The stress dependence of steady state creep

Despite many attempts to obtain a unifying interpretation of a vast range of experimental data for the variation of $\dot{\epsilon}_s$ with σ, it now seems more plausible to classify behaviour into various regimes with an appropriate model for each.[30] Historically, the most commonly used expression has taken the form $\dot{\epsilon}_s = A_1\sigma^n$, with A_1 a temperature dependent parameter and with the exponent n about 5 for many pure metals over a fairly wide range of stress. For solid solutions n is frequently about 3 and this value is usually explained on the basis of dislocation movement impeded by solute drag, though most models of creep involving lattice diffusion lead to the anticipation of such a power law.[7] More recently evidence has accumulated however to show that values of n depend significantly on creep conditions as well as on the type of material. Interpretations of the fifth power law have not yet gained wide acceptance, though the third power law would naturally become raised to the fifth power if diffusion processes were to occur predominantly along dislocation cores. It was noted from the previous section that the dislocation network for 'Class II' type behaviour with $n = 5$ is on a finer scale than that for 'Class I' behaviour with $n = 3$ and this would tend to favour dislocation core diffusion in the former case but discrepancies between predicted and measured activation enthalpies for creep may be difficult to resolve.

Dorn *et al.* have taken a broad view of the formulation of creep laws,[27] concentrating on choosing dimensionless parameters, such as A, m, n and p which can be adjusted to cover virtually all situations of steady state creep by taking

$$\dot{\epsilon}_s = A(D_e Gb/kT)(b/d)^m(\gamma_s/Gb)^p(\sigma/G)^n \tag{13}$$

Here D_e is an effective diffusion coefficient which may be equivalent to a lattice diffusion coefficient but, more generally, have additional terms taking into account, for example, dislocation core or grain boundary diffusion. These dimensionless parameters are not all independent and an extensive analysis[31] of experimental data has suggested that A and n can be interrelated through

$$\log A = -10{\cdot}5 + 3{\cdot}4n \tag{14}$$

When n takes values >3, it has also been noted that $m \sim O$, showing the negligible effect of grain size in this case. The parameter p relates to the influence of the stacking fault energy γ_s. For many fcc metals and solution solid alloys showing Class II behaviour, a value of $p \sim 3$ has been noted[32] but it has also been pointed out that increase in γ_s generally leads to decrease in n.

Whilst the value of n has often been viewed in terms of different materials, it is now widely accepted that its value is greatly influenced by the stress and temperature conditions and that a power law creep rate dependence on stress is not necessarily the best description of results. Many experimental studies, usually at relatively high stress levels suggest[5, 33] a relationship

$$\dot{\epsilon}_s = A_2\exp(B\sigma) \tag{15}$$

and linking these observations with results over a wider range of stress has led to proposals[34] that

$$\dot{\epsilon}_s = A_3(\sinh B\sigma)^n \tag{16}$$

Equation (16) can become near to the linear relation $\dot{\epsilon}_s \alpha \sigma$ at the lowest stress levels, which is widely observed in more recent tests at high sensitivity for a range of pure metals and for some alloys.[35] This has raised the question of whether there is really a continuity of creep mechanism from low to high stresses.[36] Although it would be satisfying if such continuity could be established, it is now generally recognised that such a view would be seriously misleading.[30] Following the first clear proposal by Nabarro[37] of the likelihood of a creep mechanism at low stresses, not depending on dislocation movement, but created entirely by a stress directed flux of vacancies, Cottrell quickly recognised its distinctive importance as a separately identifiable process.[2, 38]

This has become a research area of significance much greater than originally envisaged, partly because of its applicability well below the melting temperature through the role of grain boundary diffusion.[35] Equation (13) remains valid with D_e now referring to such diffusion and in this case with $m = 3$ instead of $m = 2$ for the circumstances where lattice diffusion predominates. The recent tendency to produce materials of finer grain size to improve their fracture toughness and low temperature strength has widened the role of the diffusion creep process in practice and now this deformation regime usually forms a sizeable area of the deformation mechanism maps constructed by Ashby.[30]

The essential differences between dislocation creep and creep entirely by vacancy fluxes have major consequences in practical situations.[39] In the former, stress concentrations build up at the ends of sliding interfaces, whilst in diffusion creep purely through vacancy fluxes a stress pattern[40] is established which gently rises to a maximum near the grain centres as in the idealised case in Fig. 3. The alteration of grain shape is also different in the different creep modes, as is the distribution of second phase particles, since precipitate free zones are observed near to the vacancy emitting boundaries with particle accumulation on vacancy absorbing ones.[41, 42] Because of the importance of diffusion path length, it is clear that, depending on grain shape, the resistance to diffusion creep can be markedly anisotropic.[43]

Since a switch from the predominance of the diffusion creep mode to one involving dislocation motion is anticipated on increasing stress, it is pertinent to enquire about the character and sharpness of this transition. Here a recognition of the internal stress patterns is important because, as illustrated in Fig. 4 with simple geometry, a progressive increase in applied stress would first induce dislocation movement within a small cylinder at the centre of the grain which would enlarge, adjust its shape, and at sufficiently high stress extend throughout the entire grain.[44] Such occurrences tend to make more smooth the transition between the linear stress dependence and power law creep.

A further important feature that may blur the distinction between diffusion and dislocation creep modes arises because, for many useful alloys, grain boundaries are unable to act as perfect sources and sinks for vacancies.[35] Thus diffusion creep may to some extent be inhibited. As well as having a major influence on creep resistance at low stresses, this aspect is of prime relevance to creep fracture and is considered later.

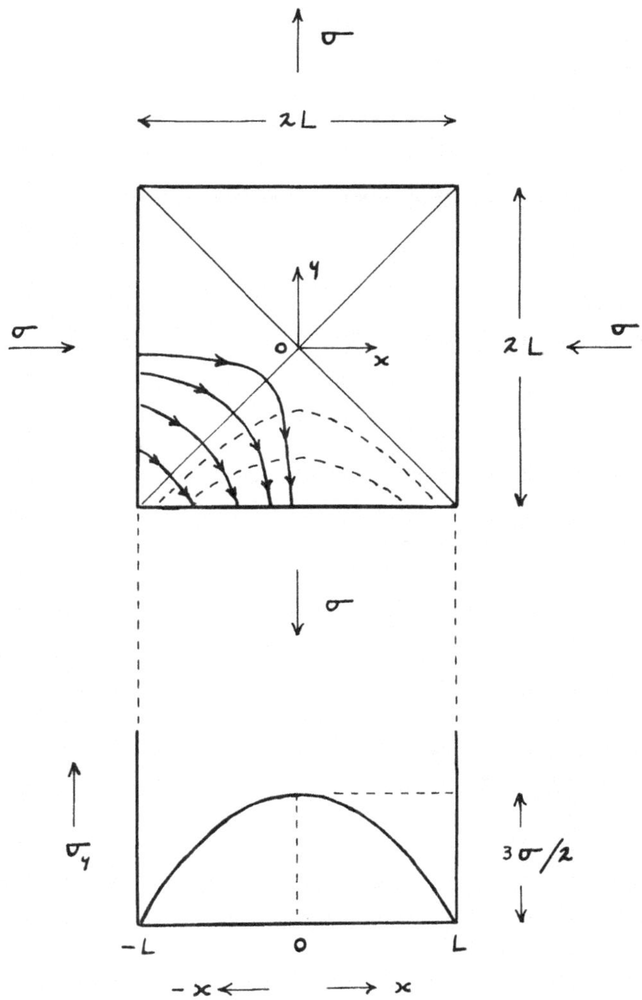

Fig. 3. When numerically equal tensile and compressive stresses act as shown on a grain of square cross section, the arrowed lines indicate the direction of the atom flux accompanied by an opposite vacancy flux to cause diffusion creep. These flux lines are hyperbolic and orthogonal to the dotted lines of vacancy concentration. The tensile stress σ_y across the horizontal faces has a parabolic form with a maximum of 3/2 times the applied stress at its centre.

The most promising approach to interpreting the inhibition of grain boundaries as vacancy sources and sinks seems to lie in the concept of discrete identification of such sources and sinks at grain boundary dislocations.[45] In pure metals these can operate with negligible impediment to the diffusion creep process and so the equations of Nabarro[37] and

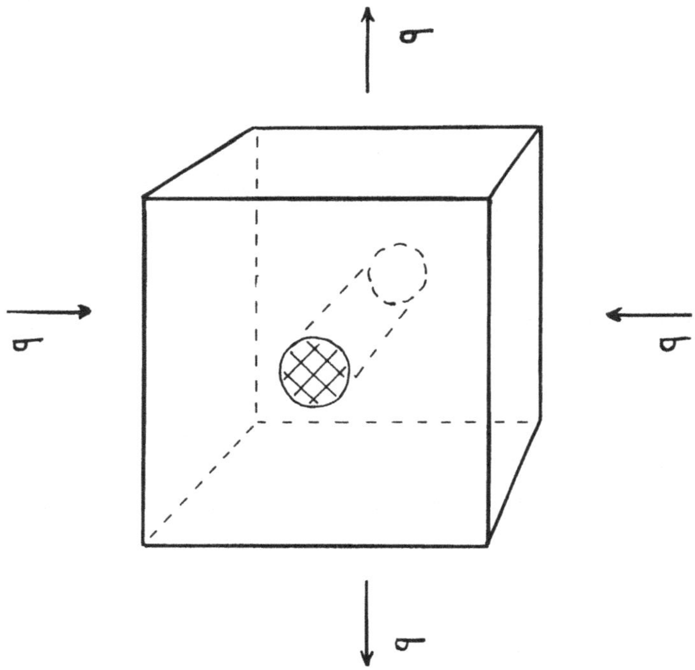

Fig. 4. In the situation illustrated in Fig. 3, surfaces of equal shear stress are cylindrical and decrease in magnitude with increasing radius. As the applied stresses are increased, dislocation movement is first initiated within such a small cylinder which grows radially, changes shape and eventually fills the entire grain.

Coble remain closely applicable apart from a small threshold stress which is necessary for grain boundary dislocations to move so that vacancy emission or absorptions can occur.

In alloys where grain boundary dislocation movement can be impeded by particle pinning or by solute drag, then there is a larger threshold stress or a finite time must be allowed for the vacancy emission and absorption processes. Diffusion creep is then said to be influenced by interface control.[7] In effect there are now two sequential steps, one involving grain boundary dislocation motion and the second the diffusion of an atom from one grain boundary to another. The resulting creep rate $\dot{\epsilon}$ can thus be written[45]

$$1/\dot{\epsilon} = (1/\dot{\epsilon}_d) + (1/\dot{\epsilon}_{\mathrm{mob}}) \tag{17}$$

where $\dot{\epsilon}_d$ is the diffusion creep rate with perfect vacancy source and sink action and $\dot{\epsilon}_{\mathrm{mob}}$ is the creep rate if it were entirely determined by grain boundary dislocation mobility, with $\dot{\epsilon}_d$ infinitely fast by comparison.

Theoretical analysis has shown[45] that $\dot{\epsilon}_{mob}$ may be expected to be proportional to σ^2/d where d is grain size, thus the purely linear variation of $\dot{\epsilon}$ with σ is destroyed. Moreover, this is a further feature that smoothes the transition between diffusion and dislocation controlled creep with increasing stress, especially in complex creep resisting alloys, but it is important that it does not obscure the distinctive characteristics of the different creep mechanisms.

The Influence of Internally Generated Stresses on Creep

It has long been known that materials may suffer deformation as a result of internal stress generation. Such manifestations occur, for example, when polycrystals comprising crystals with anisotropic thermal expansion coefficients are subject to temperature changes.[46] The question of the influence of internally generated stresses on externally stressed materials, however, was not addressed until Cottrell pointed out its relevance to the behaviour of α–Uranium used as a fuel in nuclear reactors.[47]

Here the effects of the neutron irradiation were compounded with those resulting from anisotropic crystal properties. The substantial growth (which can amount to a doubling in length in one year in a slow neutron flux of $10^{17}\,m^{-2}s^{-1}$) of single crystals of the orthorhombic α-phase in the [010] direction was attributed to the preferentially oriented deposition of the vacancies and intersticials produced by the neutron flux.[48] In randomly oriented polycrystalline uranium, which can be achieved by quenching from the β-phase, such growth is inhibited through the mutual restraints of neighbouring crystals. But, in this situation, Cottrell recognised the importance of the ability of the crystals to deform plastically if gaps were not to develop between them. The time t_e required to build up internal stress, first accommodated elastically, to cause such plastic behaviour is given by $t_e \sim \sigma_y/E\dot{\epsilon}_g$ where σ_y is the yield stress, E = Young's modulus and $\dot{\epsilon}_g$ is the irradiation induced growth rate of unrestrained crystals. The interesting question was then posed on the response of α–Uranium, in this state, to an external stress.

In proposing the answer the view was taken that the material should display Newtonian Viscosity with irradiation induced creep rate $\dot{\epsilon}_c$, under a small external stress σ, given approximately by the overall elastic strain σ/E divided by a relaxation time which could be expected to be close to the time t_e for internal plastic strain to be attained. Thus,

$$\dot{\epsilon}_c \sim \sigma/Et_e \sim (\sigma/\sigma_y)\dot{\epsilon}_g \tag{18}$$

Convincing experimental support was simultaneously provided[47] fully justifying the approach outlined above. This creep mode, as predicted, was found to result in distortion of compressively stacked α–Uranium fuel bars from the first power producing nuclear reactors and required some redesign of fuel element support for its elimination.

Other analyses of this problem have been made, including approaches[49] from the point of view of continuous deformation not explicitly involving a relaxation time. These have led to similar results, suggesting the incorporation of a numerical factor of $5/2\sqrt{3}$ in the right hand side of equation (18) which is sufficiently close to unity to demonstrate the acceptability of the different theoretical treatments.

Although, under continuous irradiation, the process described has many of the manifestations of creep at a greatly accelerated rate, it was pointed out in the original description[47] that it could more appropriately be considered as 'pseudo creep', for it was not directly related to time but rather to the magnitude and continuity of generation of the internal stress. Moreover, in principle, such creep could occur significantly at any temperature. Viewed in this light, it readily becomes apparent that this 'pseudo creep' mode of deformation may be induced by forms of internal stress generation other than irradiation.

A form of internal stress generation previously mentioned is that arising from the thermal cycling of polycrystalline materials comprising crystals with different or with anisotropic thermal expansion coefficients. Many materials fall into such a category and generally a linear external stress dependence has been found for the pseudo creep rate on thermal cycling.[50,51] Most results have been broadly in accord with the proposal that, when such materials are thermal cycled, then for each two way cycle through a large temperature interval ΔT, the increment of pseudo creep strain ϵ_i under a small applied stress σ is given by

$$\epsilon_i \sim (5/6)(\sigma/\sigma_y)\Delta\alpha[2\Delta T - \Delta T_h - \Delta T_L] \tag{19}$$

where ΔT_h and ΔT_L are respectively the temperature intervals required to attain internal yielding during the increasing and decreasing parts of the temperature cycle. The term $\Delta\alpha$ is a suitably averaged value of the difference in thermal expansion coefficients and the temperature variation of σ_y must be taken into account. The pseudo creep rate that may be developed is related only indirectly to time through the rate of change of temperature.

The analysis has been further extended[49] both to include the case where σ_y is more realistically replaced by a term I, representing the stress required to sustain the required rate of conventional creep in the material, and to deal with the situation where only small internal stresses are generated. Such a case is particularly relevant to materials subjected to small thermal cycles. It is convenient to express the results of the analysis in the form of a creep acceleration factor R representing the ratio of the rate of creep with and without thermal cycling. The time factor is now taken into account in the value ascribed to I and it is deduced that

$$R = (I/\sigma)^{n-1}[5n/(4n + 1)] \qquad (20)$$

where n is the stress exponent for conventional creep. It has been established experimentally[52] that large effects can arise in these circumstances as indicated by equation (20). Moreover, it implies that considerably enhanced creep rates may be expected for materials with anisotropic crystal structures if thermal fluctuations of only a few degrees C are continuously experienced.

Similar considerations apply to other cases where internal stresses are generated. Materials with two phases of different properties, each present in a significant volume fraction, can also be influenced by such effects and this includes a wide range of composite materials on which there remains much scope for further study. Another situation involves thermal cycling through a phase change where internal stress can be induced through the accommodation of transformed regions of volume different from that of the original phase. Experimental results[53] have clearly identified that, with application of a small external stress during such cyclic transformations, a pseudo creep process occurs which can be interpreted in somewhat similar terms to those previously described. In this case the internal stress level can be related to the flow stress σ_y of the weaker phase with the strain generated by the fractional volume difference between the two phases ($\Delta V/V$). The pseudo creep equation now takes the form

$$\epsilon_c \sim 5\sigma(\Delta V/V)/3\sigma_y \qquad (21)$$

where ϵ_c is the strain produced by each two-way cycle and with $\sigma << \sigma_y$. It is also possible to consider conventional creep leading to an internal stress I in the weaker phase, rather than ascribing a specific yield stress σ_y. In this case, it is deduced that

$$\epsilon_c \sim 4\sigma(\Delta V/V)[5n/(4n + 1)]/3I \qquad (22)$$

It is noted that equation (22) becomes identical with equation (21) as anticipated when $I \rightarrow \sigma_y$ and $n \rightarrow \infty$. Experimental results showing the linearity of pseudo creep strain at small applied stresses during phase transformations are shown in Fig. 5.

Such results have been found to have a bearing on the distortion of some alloys following heat treatment and have assisted the introduction of improved handling methods.

More generally, the observed linear variation of pseudo creep with small applied stresses implies an effective strain rate sensitivity exponent of unity whilst internal stresses are continuously generated. This is the

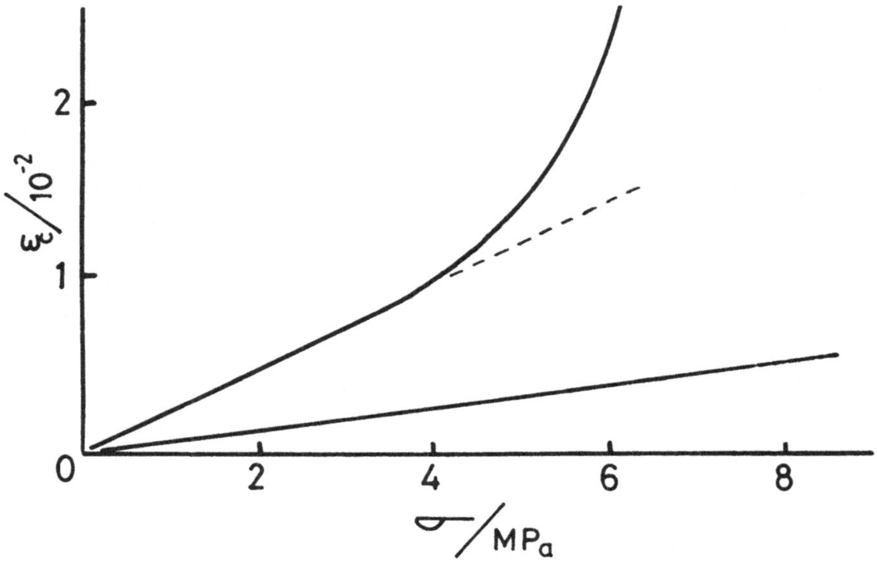

Fig. 5. The experimentally determined variation of strain ϵ_c generated under external stress σ for each two way cycle through a phase transformation.[53] *The upper curve is for the α–γ transformation in iron at 910 °C and the lower curve is for the $\alpha + Fe_3C - \gamma$ transformation in an iron – 0·39mass percent carbon alloy.*

condition for superplastic behaviour[54] whereby considerable tensile deformation can be sustained without neck formation.[55] A significant research area has developed from this with the particular phenomenon now known as internal stress superplasticity,[56] which may be utilised, for example, in the manufacture of components of complex shape from material in sheet form.

Creep Fracture

One of the many areas where Alan Cottrell brought new insight into the basis of old ideas concerns the long held view that grain boundaries strengthen metals at low temperatures but weaken them above some higher 'equicohesive' temperature.[57] This view may relate to fracture as well as to plasticity and one example is the fracture that may occur in many materials after quite small strain caused by a low tensile stress at elevated temperature.

In almost all cases of creep fracture, separation is known to occur at grain boundaries and primarily on those boundaries oriented nearly perpendicular to the principal tensile stress.[58,59] It can arise by crack growth originating from triple points where three grain boundaries meet or from cavity formation along the grain boundaries or from both these causes. There are distinctive features about these processes which will first be considered separately.

Cracking at triple points

A valuable approach to this form of cracking has resulted from the consideration of the stress concentration built up by the intersection of sliding at two grain boundaries subject to high shear stress in causing a crack to form at the third grain boundary across which there is a high tensile stress. Such stress concentrations can be sufficiently severe to cause fracture and there have been several analyses of this situation following the features that Zener[60] previously emphasised. This remains a complex research area with some models put forward where such cracks grow continuously[61, 62] and others which involve cavity nucleation at the head of a crack[63] and propagation by their link up.

Not all metals behave in a similar way. For example, in pure Aluminium such cracking has not been observed at temperatures where extensive grain boundary sliding occurs. One general observation, however, reported for a wide variety of metals[7] is that cracking is inhibited and ductility greatly enhanced if some grain boundary migration takes place because the stress concentration is then effectively dispersed.

When triple point cracking does occur, plausible interpretations have been derived[64, 65] from analogy with brittle cleavage fracture at low temperature by adaptation of the theory of cleavage crack formation by the coalescence of dislocations slipping on intersecting planes.[66] Following the intersection of sliding grain boundaries at a rate $\dot{\epsilon}_b$, it has been deduced that in a material of grain size d under a tensile stress σ, a crack will become unstable and will propagate rapidly to cause failure after a time t_f given by

$$t_f \sim 2\gamma/\mathrm{d}\dot{\epsilon}_b\sigma \tag{23}$$

where γ is the energy required to separate unit area of fracture surface. As in so many other fracture problems, there remains the difficulty of assigning a value to γ because of energy consuming processes additional to the energy requirement of breaking atomic bonds and so equation (23) has only limited predictive capability. Further, the derivation of equation (23) presumes that the crack becomes unstable before traversing the complete grain boundary. When it is more appropriate to consider entire grain boundary separation before fracture, then the failure criterion is based on simultaneous but initially independent crack development in neighbouring grains. This leads to the prediction of fracture after a time.

$$t_f \sim (4/\dot{\epsilon}_b)(\gamma/Gd)^{1/2} \tag{24}$$

It is noted that stress does not appear directly in equation (24) but its importance lies in the dependence of $\dot{\epsilon}_b$ on σ. Although differing in the form of relationship, both the above equations show that finer grain size material is less prone to short term fracture at elevated temperatures. This aspect is now well bourne out in practice whereby limitation of austenite grain size during the welding of steam pipe line materials has proved to be a crucial factor in improving their lifetime in service.

If the grain boundary sliding rate $\dot{\epsilon}_b$ matches or is proportional to the overall creep rate $\dot{\epsilon}$, then it is noted that equation (24) provides support for the approximate but widely applicable Monkman–Grant relationship that $\dot{\epsilon}t_f$ is nearly constant for a given material over an extensive range of stress and temperature.[67] The validity of this relationship for long creep life is more questionable and modified relationships have been proposed.[68] It is generally recognised that failure mechanisms depending primarily on grain boundary shear processes are most appropriate to interpret fracture in strong materials under high stresses which cause failure in a relatively short time.[69]

Cavitation

Particularly at rather low stresses, a quite different fracture mode is observed which was first characterised by the detection of nearly spheroidal cavities typically spaced a few microns apart mainly on grain boundaries subjected to a perpendicular tensile stress.[70] This phenomenon is termed 'cavitation' and it remains a subject of detailed study. Cavities can form copiously from an early stage of creep and creep life is then determined by their nucleation and growth until the cross section of the material is severely reduced.

Cavity nucleation

As with so many other instances of nucleation phenomena across broad areas of science, the mechanism of cavity nucleation continues to prove difficult to resolve in detail. Nucleation is often aided by the presence of particles on the grain boundaries with poor adhesion to the matrix. Gas bubbles above a critical size, and the segregation of certain elements can also assist the nucleation process. In steels, attempts[71] have been made to list a heirarchy of particles in relation to their effectiveness as cavity nucleants. Such endeavours are far from complete but suggest that cavities are most easily nucleated by sulphides and with nucleation on oxides enhanced by the presence of sulphur. Carbides appear to have a less important role, though some results are in direct conflict with this.[72] It is likely that carbide composition, size, morphology and orientation all need to be taken into account. It is generally expected that, if significant stress concentrations are not produced, the size of nucleants is likely to lie within the range 0·03–0·5 μm, depending on the material and on the stress level. This is a difficult size range for *in situ* investigation, being below the resolution limit of optical microscopy but too large for transmission microscopy except for the high voltage instruments. The rarity of the nucleation events adds further difficulty, though observations have been reported on thin films of initial stages of the nucleation process.[73, 74] An alternative technique, again with some limitations, has been developed through the subsequent fracturing of the material at low temperatures along the cavitated boundaries.[75] This is accomplished for iron with low carbon content but with the introduction of a small quantity of oxygen which makes low temperature intergranular fracture possible. An example is shown in Fig. 6. Not many materials involved in cavitation studies can be induced to fracture in this way at least until the level of cavitation is severe and so the technique is of limited application. More generally, the

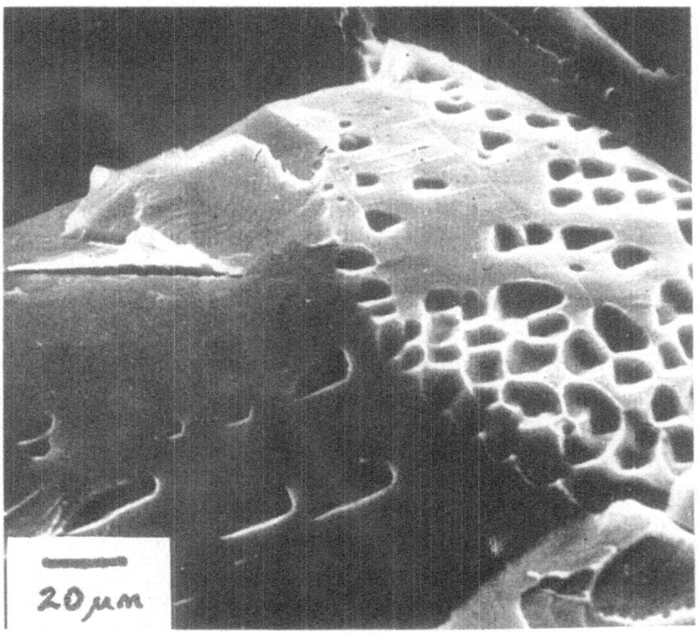

Fig. 6. A scanning electron micrograph of cavities on grain boundaries in iron after creep. A small oxygen content in the iron permits intergranular fracture on impact at low temperatures so that cavitation can be followed at various stages of creep. Note that cavities are facetted and are not uniformly distributed.
(B. J. Cane.)

scanning electron microscope examination of carefully prepared sections continues to prove a useful technique.

Many proposals have been made about nucleation mechanisms with early studies extensively reviewed.[76,77] Rate theories of cavity nucleation lead to the expectancy of strong dependencies of nucleation rate on stress and on temperature. Although predictions are substantially influenced by cavity shapes,[78] it remains difficult to provide experimental support. A relatively common observation has been the nearly linear increase of the number of cavities with strain in the early stages of creep,[79,80] suggesting that strain rather than time is a critical parameter. Moreover, cavities have not always been found to be associated with particles though it is not easy to obtain definitive information on this point. Thus cavity nucleation models involving grain boundary sliding or the interaction of such sliding with grain deformation may be relevant to the nucleation process.[81] In some Nimonic alloys with grain boundary particles it has been shown that cavity nucleation can be greatly enhanced by prior strain at room temperature, again emphasising the importance of deformation.[74] It is generally agreed that relatively large sliding displacements are needed for cavity nucleation[82] and this may be linked with the localisation and irregularity of such displacements.[83]

Cavity growth

In contrast with nucleation, the mechanisms of cavity growth are much better understood but it is instructive first to recall the reasoning that has led to a broader interpretation of fracture processes and to a clear distinction between the cases where they occur with and without stress concentrations.

Following Griffith, the minimum tensile stress to grow an internal crack of length $2c$ is given approximately by $(E\gamma/c)^{1/2}$ where E is Young's modulus and γ the energy to separate unit area of the fracture surfaces. At a grain boundary E and γ are typically each reduced by about 25% below their value within the grain, but this is insufficient to cause serious reduction in fracture strength unless γ is substantially decreased by segregation. Further, the Griffith condition represents a minimum and necessary, but not sufficient, requirement for fracture. It is also necessary that the crack tip be sufficiently sharp for the stress concentration to reach a level $\sim E/10$ where atomic bonds can be broken.

An entirely different situation pertains for the growth of grain boundary cavities at temperatures high enough for diffusive processes to operate. In this case thermal agitation greatly assists in breaking atomic bonds and so high stress concentrations are not required. Providing a cavity is of sufficient size, a small applied stress is adequate for its growth. This

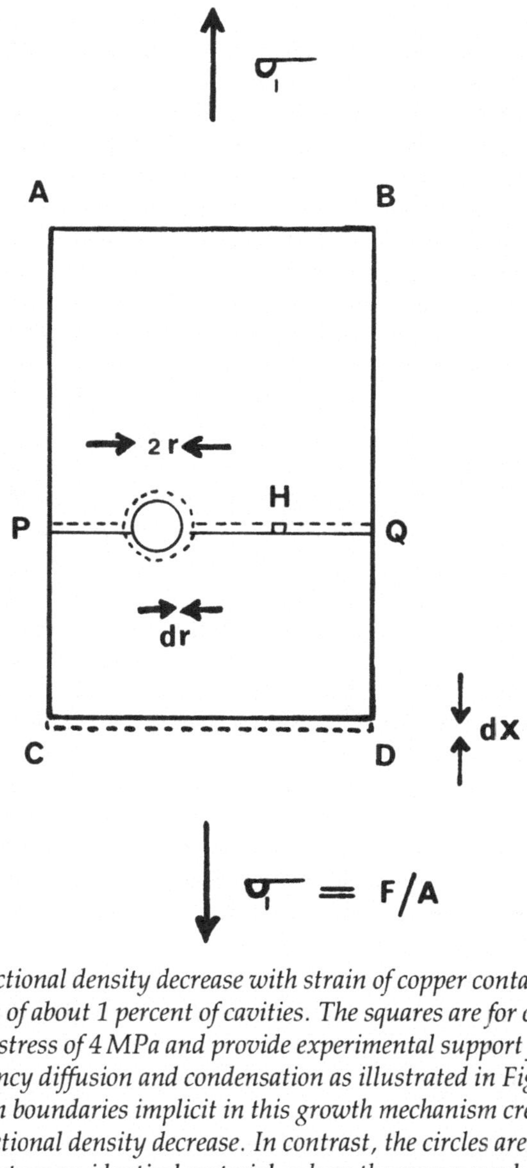

Fig. 7. The fractional density decrease with strain of copper containing an initial volume fraction of about 1 percent of cavities. The squares are for creep strain at 500 °C under a stress of 4 MPa and provide experimental support for cavity growth by vacancy diffusion and condensation as illustrated in Fig. 7. The atom plating on grain boundaries implicit in this growth mechanism creates a strain equal to the fractional density decrease. In contrast, the circles are for creep strain at room temperature on identical material, where the copper work hardens and diffusion is negligible. In this case the density decrease shows that the increase in cavity volume divided by their initial volume is approximately equal to the strain, supporting the theory of hole growth by plastic deformation.
(M. D. Hanna.)

situation is illustrated in Fig. 7 where the cavity grows by acquiring vacancies, created at and diffusing along the grain boundary under tension, to increase the cavity radius by dr so that the surface energy is increased by $4\pi\gamma[(r + dr)^2 - r^2]$ which is approximately $8\pi\gamma r dr$. Now the diffusion of vacancies from the grain boundary implies an effective flow of atoms in the opposite direction which plate on the boundary[84] and this will cause an extension x in the direction of the force given by $x = 4\pi r^2 dr/A$. The force has now done work σAx which is sufficient to enlarge the cavity when it exceeds $8\pi\gamma r dr$. It immediately follows that cavity growth by this mechanism requires $\sigma > 2\gamma/r$. When the dihedral angle at the cavity intersection with the grain boundary is taken into account, the value of r is then the radius of each spherical section of the surface of the cavity, but in practice this makes little difference. The analysis also applies when cavities are facetted through considering an equivalent radius.

The above discussion is a prime example on the importance of the two criteria for all forms of fracture which Cottrell so clearly emphasised;[2] namely, the thermodynamic condition must be met whereby the energy supplied must exceed that for interfacial separation and secondly there must be some atomic process by which the mechanism of fracture can operate. It is essential to appreciate that these criteria can be met in different ways. This has not always been sufficiently recognised for many doubts have been cast on the ability of cavities to grow by vacancy diffusion and condensation.[76] It is only recently that some of the reasons for confusion have been resolved.

As well as experimental demonstrations of the effectiveness of grain boundaries in pure metals as sources and sinks for vacancies in diffusion creep experiments,[36] much additional supporting evidence has accumulated from sintering experiments and from Helium bubble growth on annealing after α-particle bombardment.[85] Its relevance to cavity growth was first clearly demonstrated by the experiments of Hull and Rimmer[86] on the creep of copper under hydrostatic pressure where cavity growth was entirely inhibited when such pressure was equal to the tensile creep stress. Arguments pointing out discrepancies between the theory of cavity growth by vacancy diffusion and experimental measurements of total cavity volume and of creep life were largely overcome by taking into account cavity nucleation with strain.[87]

Perhaps the most direct evidence for cavity growth by vacancy diffusion is derived from the situation where cavities are already nucleated above the critical size in relation to the subsequent creep stress which can be too low to induce further nucleation. One method of achieving this prior nucleation is to diffuse tritium at high pressure which decays to form helium which agglomerates into bubbles on heat treatment.[88] Another method[89] has involved exposing copper containing an appropri-

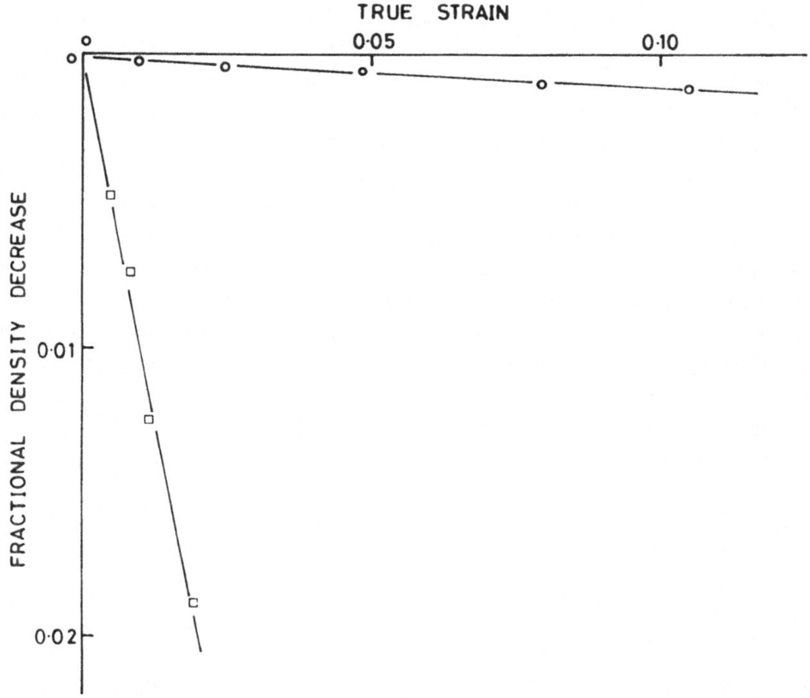

Fig. 8. A schematic illustration of the growth of a cavity of radius r *and surface energy* γ *per unit area by the condensation of vacancies originating in the grain boundary PQ and diffusing along it under the stress* σ_1*. This process effectively removes atoms from the cavity surface to cause atom plating at the grain boundary and an increase in length in the direction of the force* $F = \sigma_1 A$ *where* A *is the area of cross section. The work done is sufficient to provide the increased surface energy of the cavity when* $\sigma_1 > 2\gamma/r$.

ate distribution of cuprous oxide particles to a partial pressure of hydrogen at about 550 °C. The oxide particles are then reduced to form steam bubbles. When material initially containing such bubbles, each about 5 μm in diameter, is subject to creep at 500 °C at stresses as low as 4 MPa then the vacancy growth relationships are closely followed without the complications of further nucleation and the increase in total cavity volume divided by the specimen volume is nearly equal to the creep strain.[89] This is fully in accord with the creep strain derived from atom plating on grain boundaries which must necessarily accompany the vacancy diffusion growth mechanism. In contrast, if identical material is, instead, strained at room temperature the increase in cavity volume divided by the initial cavity volume is approximately equal to the strain as anticipated in the theory of hole growth by dislocation related plastic deformation processes.[90] The striking experimentally determined difference in the relation between cavity growth rates in the two circumstances is shown in Fig. 8.

These clear observations have far reaching significance in practice. For example, in the simple schematic diagram in Fig. 9, for cavity growth by vacancy diffusion, only those cavities on the grain boundary CD grow under stress σ_1 whilst those on AB only respond to σ_2. Failure is thus primarily governed by the principal tensile stresses in this case, with each stress having essentially an independent effect. The case for dislocation related cavity growth by plastic deformation is quite different, with such growth then related to an effective Von Mises stress σ_e given by $\{[(\sigma_1 - \sigma_2)^2 + (\sigma_2 - \sigma_3)^2 + (\sigma_3 - \sigma_1)^2]/2\}.^{1/2}$ These considerations are of crucial importance in the interpretation of notch effects in high temperature deformation and to an overall interpretation of the influence of multiaxial stresses.[91]

In the foregoing presentation emphasis has been placed on clear distinctions between the two primary cavity growth processes and failure modes, one by vacancy diffusion and the other involving dislocation movements, because of the importance of recognising their individual characteristics. It is increasingly appreciated that in some intermediate stress range they may operate simultaneously and have interacting effects. This has formed a large research area with much attention given to modes of coupling between the two processes.[92, 93] It is envisaged that cavities may sometimes grow by acquisition of vacancies only from limited regions of the grain boundary immediately surrounding them whilst elsewhere the material is creeping through dislocation movements. Further considerations have also taken into account non uniform cavity distributions whereby their growth may be constrained by insufficient plastic deformation in neighbouring grains.[94]

A highly significant additional factor, which is fortunate for the successful application of many alloys for service under stress at high temperatures, is that their grain boundaries are unable to act as perfect sources for vacancies. As mentioned earlier, alloying can often, though not always, modify grain boundary behaviour in this respect.[35] Here some problems of cavity growth become closely related to those of diffusion creep controlled by vacancy emission and absorption. The progress made in understanding diffusion creep in complex alloys is thus of particular value.[45, 95]

Gradually, it is becoming possible to interpret creep fracture in an increasingly wide range of practical situations. Areas where a particular mechanism may predominate have been usefully depicted on 'Fracture Maps',[96] but it is emphasised that such maps may change greatly in form

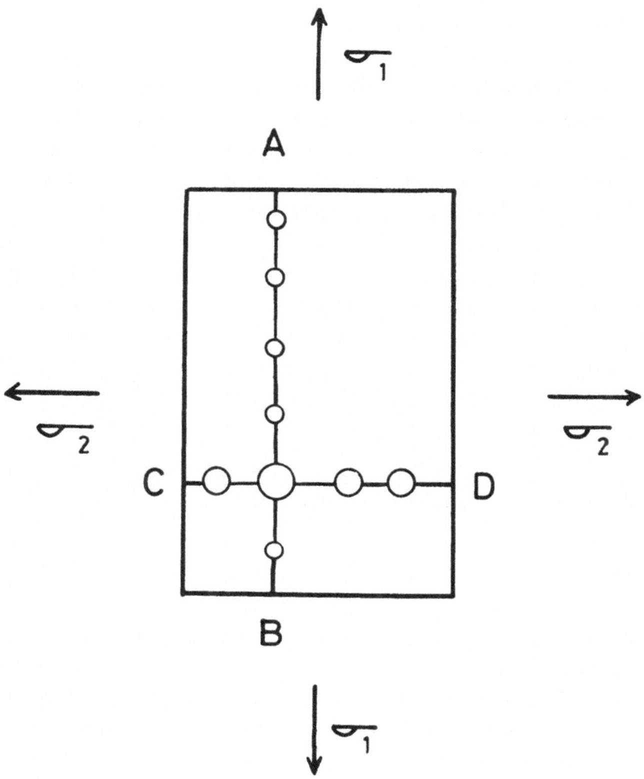

Fig. 9. A schematic illustration of cavity growth by vacancy diffusion and condensation under biaxial stress, where only the cavities on the grain boundary CD grow under the influence of the stress σ_1 and likewise cavities on AB grow under stress σ_2. Cavity growth in this case is strongly sensitive to the orientation of the grain boundaries on which they are situated and the principal stresses have independent effects. In contrast, cavity growth through plastic deformation is governed by the Von Mises stress, not by individual principal stresses, and is not dependent on grain boundaries.

even with slight changes in composition or microstructure in nominally similar materials. Attention given to the ways in which different mechanisms may interact, particularly for cavity growth, leads to the recognition[97] of sequential and parallel processes and models developed along these lines show much promise in evaluating the range and validity of many empirical rules.

Conclusions

Alan Cottrell's perceptions of the mechanisms of creep and fracture, mainly expounded between 25 and 40 years ago, have not only remained valid but have often been a stimulus for much subsequent work. Perhaps the most important feature, running throughout his contributions, lies in the breadth of approach, rarely drawing upon creep tests in isolation but frequently bringing together ideas and information from other areas.

It is now clear, for example, that the concepts of solute drag on dislocation movement at elevated temperatures, the form of creep transients following stress changes, the role of internally generated stresses causing pseudo creep, the ability of grain boundaries to act as vacancy sources and sinks and the conditions where vacancy condensation plays a part in creep fracture have all been highly significant in assisting further developments and understanding. Though the picture is incomplete, the number of puzzling features has been greatly reduced.

References

1. A. J. Kennedy, *Processes of Creep and Fatigue in Metals*. Oliver and Boyd, Edinburgh, 1962.
2. A. H. Cottrell, *The Mechanical Properties of Matter*. Wiley, New York, 1964.
3. E. N. da C. Andrade, *Proc. Roy. Soc. London*. **A84**, 1, 1911.
4. A. H. Cottrell and V. Aytekin, *Nature*, **160**, 328, 1947.
5. A. H. Cottrell and V. Aytekin, *J. Inst. Met.* **77**, 389, 1950.
6. L. N. G. Filon and H. T. Jessop, *Phil. Trans. Roy. Soc.* **223A**, 80, 1928.
7. J. Cadek, *Creep in Metallic Materials*. Elsevier, Amsterdam, 1988.
8. A. Graham and K. F. K. Walles, *J. Iron & Steel Inst.* **179**, 105, 1955.
9. A. H. Cottrell, *J. Mech. & Phys. Solids* **1**, 53, 1952.
10. F. Dobes and J. Cadek, *Met. Mater.* **19**, 31, 1981.
11. R. W. Evans and B. Wilshire, *Creep of Metals and Alloys*. Institute of Metals, London, 1985.
12. R. W. Evans, J. D. Parker and B. Wilshire, *Recent Advances in Creep and Fracture of Engineering Materials and Structures*, (ed. B. Wilshire and D. R. J. Owen). Pineridge Press, Swansea, p. 135, 1982.
13. G. W. Greenwood, 'The Prediction of Creep Life', Proceedings of International Conference on Residual Life of Power Plant Equipment — Prediction and Extension', Bharat Heavy Electricals Ltd, Hyderabad, India, 1989.
14. R. W. Bailey, *J. Inst. Met.* **35**, 27, 1926.
15. E. Orowan, *J. West of Scotland Iron & Steel Inst.* **54**, 45, 1946–7.
16. J. D. Parker and B. Wilshire, *Met. Sci.* **12**, 453, 1978.
17. W. Blum, J. Hausselt and G. Konnig, *Acta. Met.* **24**, 293, 1976.
18. J. P. Poirier, *Acta Met.* **25**, 913, 1977.
19. B. Burton, *Czech. J. Phys.* **B38**, 505, 1988.
20. D. McLean, *Trans. AIME* **242**, 1193, 1968.

21. S. Takeuchi and A. S. Argon, *J. Mater. Sci.* **11**, 1542, 1976.
22. J. Weertman, *J. Appl. Phys.* **28**, 1185, 1957.
23. J. P. Hirth and J. Lothe, *Theory of Dislocations*, McGraw Hill, New York, 1968.
24. A. H. Cottrell and M. A. Jaswon, *Proc. Roy. Soc.* **A199**, 104, 1949.
25. J. Friedel, *Dislocations*. Pergamon Press, Oxford, 1964.
26. A. K. Mukherjee, J. E. Bird and J. E. Dorn, *Trans. ASM* **62**, 155, 1969.
27. J. E. Bird, A. K. Mukherjee and J. E. Dorn, *Quantitative Relationships between Properties and Microstructure* (Ed. D. G. Brondon and A. Rosen), Israel Univ. Press, Jerusalem, p. 255, 1969.
28. O. D. Sherby and P. M. Burke, *Prog. Mater. Sci.* **13**, 325, 1968.
29. F. A. Mohamed and T. G. Langdon, *Acta Met.* **22**, 779, 1974.
30. M. F. Ashby, *Acta Met.* **20**, 887, 1972.
31. R. L. Stocker and M. F. Ashby, *Scr. Metall.* **7**, 115, 1973.
32. A. Orlova, *Met. Mater.* **7**, 242, 1969.
33. F. Garofalo, *Fundamentals of Creep and Creep–Rupture in Metals*. MacMillan, New York, 1965.
34. G. Garofalo, *Trans AIME* **227**, 351, 1963.
35. B. Burton, 'Diffusional Creep in Polycrystalline Materials', Trans. Tech. Publications, Aedermannsdorf, Switzerland, 1977.
36. B. Burton and G. W. Greenwood, *Acta Met.* **18**, 1237, 1970.
37. F. R. N. Nabarro, Report on Conference on the Strength of Solids, Phys. Soc., London, p. 75, 1948.
38. A. H. Cottrell, *Dislocations and Plastic Flow in Crystals*, Oxford, 1953.
39. G. W. Greenwood, 'Diffusion Creep and its Technological Relevance', in *Physical Metallurgy of Reactor Fuel Elements* (Ed. J. E. Harris and E. C. Sykes). The Metals Society, London, 1975.
40. G. W. Greenwood, *Fundamentals of Deformation and Fracture* (Eds B. A. Bilby, K. J. Miller and J. R. Willis). Cambridge Univ. Press, p. 313, 1985.
41. R. C. Squires, R. T. Weiner and M. Phillips, *J. Nucl. Mat.* **8**, 77, 1963.
42. J. E. Harris and R. B. Jones, *J. Nucl. Mat.* **10**, 360, 1963.
43. G. W. Greenwood, *Phil. Mag.* **A51**, 537, 1985.
44. G. W. Greenwood, *Res. Mechanica.* **14**, 61, 1985.
45. E. Arzt, M. F. Ashby and R. A. Verrall, *Acta Met.* **31**, 1977, 1983.
46. W. Boas and R. W. K. Honeycombe, *Proc. Roy. Soc.* **186A**, 57, 1946.
47. A. C. Roberts and A. H. Cottrell, *Phil. Mag.* **1**, 711, 1956.
48. A. H. Cottrell, *Met. Rev.* **1**, 479, 1956.
49. R. G. Anderson and J. F. W. Bishop, Symposium on 'Uranium and Graphite', Inst. of Metals, London, Paper No. 3, 1963.
50. W. S. Blackburn, G. Harnby and J. J. Stobo, *J. Nucl. Energy*, **12**, 162, 1960.
51. V. A. Likhayer, G. A. Malygin, A. V. Nikiforov and G. A. Vladimirov, *Fiz. Met. Metalloved* **16**, 908, 1963.
52. L. R. T. Gardner and W. N. Miller, Symposium on 'Uranium and Graphite', Inst. of Metals, London, Paper No. 4, 1962.
53. G. W. Greenwood and R. H. Johnson, *Proc. Roy. Soc.* **283A**, 403, 1965.
54. J. Pilling and N. Ridley, *Superplasticity in Crystalline Solids*. Inst. of Metals, London, 1989.
55. R. C. Lobb, E. C. Sykes and R. H. Johnson, *Met. Sci. J.* **6**, 33, 1972.

56. O. D. Sherby, *ISIJ International* **29**, 698, 1989.
57. A. H. Cottrell, 'Intercrystalline Creep Fractures' in Structural Processes in Creep, Special Report No. 70, Iron and Steel Inst, London, p. 1, 1961.
58. J. H. Gittus, *Creep Viscoelasticity and Creep Fracture*. Elsevier, Applied Science Publishers, London, 1975.
59. H. E. Evans, *Mechanisms of Creep Fracture*. Elsevier, Applied Science Publishers Ltd, Barking, UK, 1984.
60. C. Zener, *Fracturing of Metals*. ASM Cleveland, Ohio, USA, 1948.
61. W. B. Beere and M. V. Speight, *Met. Sci.* **12**, 593, 1978.
62. D. S. Wilkinson, *Mater. Sci. & Eng.* **49**, 31, 1981.
63. T. Hasegawa and B. Ilschner, *Acta Met.* **33**, 1151, 1985.
64. J. A. Williams, *Acta Met.* **15**, 1559, 1967.
65. J. A. Williams, *Phil. Mag.* **15**, 1289, 1967.
66. A. H. Cottrell, *Trans AIME* **212**, 192, 1958.
67. F. C. Monkman and N. J. Grant, *Proc. ASTM* **56**, 593, 1956.
68. F. Dobes and K. Milicka, *Met. Sci.* **15**, 186, 1977.
69. D. McLean, *Mechanical Properties of Metals*. Wiley, New York, 1964.
70. J. N. Greenwood, D. R. Miller and J. W. Suiter, *Acta Met.* **2**, 250, 1954.
71. E. P. George and D. P. Pope, 7th International Symposium on Creep–Resistant Metallic Materials, Central Council of the Czech Scientific & Technical Soc. Brno, Czechoslovakia, 1986.
72. I. W. Chen and A. S. Argon, *Acta Met.* **29**, 1321, 1981.
73. R. G. Fleck, D. M. R. Taplin and C. J. Beevers, *Acta Met.* **23**, 415, 1975.
74. B. F. Dyson, *Scripta Met.* **17**, 31, 1983.
75. B. J. Cane and G. W. Greenwood, *Met. Sci.* **9**, 55, 1975.
76. A. J. Perry, *J. Mater. Sci.* **9**, 1016, 1974.
77. D. McLean, B. F. Dyson and D. M. R. Taplin, 4th Int. Conf. on Fracture (Ed. D. M. R. Taplin), **1**, 325, Waterloo, Canada, 1977.
78. A. S. Argon, *Recent Advances in Creep and Fracture of Engineering Materials and Structures* (Ed. B. Wilshire and D. R. J. Owen). Pineridge Press, Swansea, p. 1, 1982.
79. G. W. Greenwood, *Phil. Mag.* **19**, 423, 1969.
80. D. A. Woodford, *Met. Sci.* **3**, 50, 1969.
81. R. C. Gifkins, *Acta Met.* **4**, 98, 1956.
82. W. D. Nix, *Scripta Met.* **17**, 1, 1983.
83. A. S. Argon, *Scripta Met.* **17**, 5, 1983.
84. J. E. Harris, M. O. Tucker and G. W. Greenwood, *Met. Sci.* **8**, 313, 1974.
85. R. S. Barnes, G. B. Redding and A. H. Cottrell, *Phil. Mag.* **3**, 97, 1958.
86. D. Hull and D. E. Rimmer, *Phil. Mag.* **4**, 673, 1959.
87. G. W. Greenwood, *Phil. Mag.* **8**, 707, 1963.
88. M. D. Hanna and G. W. Greenwood, *Acta Met.* **30**, 719, 1982.
89. J. M. Mintz and A. K. Mukherjee, 3rd Int. Conf. on Creep and Fracture of Engineering Materials and Structures (Ed. B. Wilshire and D. R. J. Owen), Inst. of Metals, London, 1987.
90. J. W. Hancock, *Met. Sci.* **10**, 319, 1976.
91. F. A. Leckie, *Phil. Trans. Roy. Soc.* **288A**, 27, 1978.
92. W. B. Beere and M. V. Speight, *Met. Sci.* **12**, 593, 1978.

93. G. H. Edward and M. F. Ashby, *Acta Met.* **27**, 1505, 1979.
94. B. F. Dyson, *Met. Sci.* **10**, 349, 1976.
95. S. J. Dobson, H. Jones and G. W. Greenwood, 'Creep of Stabilised Stainless Steel at Low Stresses', in *Perspectives in Metallurgical Development*. Metals Soc, London, p. 291, 1984.
96. M. F. Ashby, C. Gandhi and D. M. R. Taplin, *Acta Met.* **27**, 729, 1979.
97. W. B. Beere, *Cracks and Cavities in Creep and Fatigue* (Ed. J. H. Gittus). Applied Science Publishers, London, p. 29, 1981.

10 Fibre Reinforcement at the Metallurgy Laboratory at Cambridge 1958–67

A. Kelly

Vice-Chancellor, University of Surrey

Abstract
I am pleased to be invited to make this presentation since I have for many years admired Sir Alan Cottrell's work as a scientist and he has been extraordinarily helpful and encouraging to me personally.

The background to the commencement of work on strong fibrous solids in the Cambridge Metallurgy Laboratory is, as I remember it, as follows.

1. Introduction

Cottrell had just been appointed Professor in 1958 and he offered me a job as a Lecturer and asked me (I knew very little of Metallurgy at the time) to give a set of lectures on ceramics. I came to Metallurgy at Cambridge with the first Instron imported into Britain and from a Department at Northwestern University which, under Morrie Fine's leadership, had already discovered the importance of materials science as a discipline, of which metallurgy was to be seen as a part. I set about preparing these lectures and remember that I did this by reading Slater's Quantum Theory of Matter[1] while crossing the Atlantic by ship. Other people lecturing on ceramics would have scoffed at that but, as it happened, it stood me in very good stead. I discovered, while preparing the lectures, rather little science, or rather, little materials science of ceramics. There was a wealth of chemistry but very little material relating structure to properties. I persevered and discussed with the undergraduates things like the physical interpretation of Mohs hardness scale, the difficulty of moving dislocations in materials such as sapphire, why silicon carbide sintered slowly and, following Slater, why covalent solids are stiff and of low density, since the atoms are held apart by spatially directed bonds. We discussed a host of other things not usually dealt with in ceramics courses. I was pleased afterwards to discover that the graduates who went into such industries as the steel industry found my lectures very useful, even though I had not dealt in excruciating detail with how furnace linings were rammed into place.

Despite the wealth of material to read, it is difficult to appreciate nowadays that some very obvious data was missing, e.g. the elastic constants of boron, or even of fully dense graphite. At Cambridge I started some work on ceramics, notably, for what comes later, on the investigation of graphite single crystals and on magnesium oxide. My own special interest was in precipitation hardening of metals and I discovered that, in aluminium–copper and copper–beryllium and suchlike, precipitation hardened alloys in the overaged condition developed very strong rates of work-hardening. We managed to show that the yield stress was controlled by the Orowan stress.

2. Cottrell's Statement

AHC, at the time, was becoming a very fashionable young scientist and so was invited to give lectures such as the Royal Society's Bakerian, and discourses to the Royal Institution, Royal Society of Arts, etc. He often used to give me the text of these lectures to read before delivering them.

In a discourse delivered to the Royal Institution on 15 June 1960 he enunciated the principle of fibre reinforcement.[2] I quote what he said:

> No, the practical approach is to admit the existence of cracks and notches and to try to render them innocuous. Suppose that we are stretching a rod . . . and . . . that it consists of a bundle of parallel . . . fibres joined together — by some adhesive or solder. If there is a transverse notch in the rod, cutting across a number of these fibres, the forces from the cut fibres can be transmitted to the fibres at the tip only by passing as shearing forces through the layers of adhesive . . . If this adhesive has a fairly low resistance to shear . . . it will then be incapable of focussing the transmitted forces sharply . . . There are tremendous possibilities for developing this principle further, particularly by using fibres of materials with very strong atomic forces, e.g. refractory oxides and carbides.

It is all there.

The discourses of the Royal Institution give no references, or rather AHC gave no references, to that particular talk. 'Had he dreamt the principle up for himself?', was the question I asked. 'Was there some literature unknown to me?' The hint that there was an extensive literature occurred in the list of exhibits in the library, listed at the end of the lecture.

These were from the Royal Aircraft Establishment. People there had approached strong materials in a completely different fashion from that of the metallurgist. Cottrell will have learnt of this from his membership of various quasi-confidential Government committees which abounded in that period and on which most academics interested in physical metallurgy sat. The situation of composite science at that time in Britain, and its subsequent development, is well described by McMullen.[11]

These people approached the idea of strong solids in a completely different way than did the metallurgist. This was because, I believe, at the time, the metallurgist was beginning to understand — due principally, in my opinion, to the work of George Irwin at the Naval Research Lab — the concepts of toughness. Strength for its own sake was discounted. It was known how to make high-strength steels but these were very brittle. The exponents of fibre-reinforcement — including Cottrell — had no answer to the statement or accusation that the material would lack ductility and hence would be brittle and useless. Clever answers like 'The razor blade is a very brittle material, and yet you use that', did not carry much conviction.

Because of my interest and knowledge in ceramics I had no difficulty in convincing myself that the ceramic materials, aluminium oxide, the carbides, etc., were ideal materials for blocking dislocations in motion in a metal and we had, as I say, discovered high rates of work hardening of precipitation hardened materials in the over-aged condition. I also knew, from reading Slater, why strong and light solids would be bound by directed covalent bonds, hence stiff and of low density with high melting points, small coefficients of thermal expansion etc.

There was another strand in the thread of the argument for strong fibrous materials and that had been the discovery of whisker crystals in the mid-1950s. These were very strong. The difficulties of growing a large number reproducibly, harvesting them, arranging them in a sensible fashion and making a sizeable piece, were enormous. Nonetheless the defence agencies were prepared to spend money in that area.

What had produced such an interest in fibre reinforcement in the United States in the late 1950s had been the development of the Polaris missile system, which used a glass reinforced plastic motor case, or rather filament wound motor case, and the discovery of a strong and very stiff fibrous form of boron by Talley, published in 1959.[3] I do not know whether Cottrell knew of that. He did not mention it in the RI lecture.

People concerned with fibre reinforcement or the use of whisker crystals as structural materials and those concerned with high strength metals, which were the dominant strong materials at the time, were not really able to talk to one another. The division in thinking between those two groups has persisted in some measure to the present day.

3. Pull-out

I and my group of research students were very fortunately placed in that we were elucidating the principles of the strengthening of metals and yet were lucky enough to spot the advantages of non-metallic solids as the major strengthening agents, because I was lecturing on ceramics.

In the summer of 1961 Phillip Bowden, Cottrell and others organised a closed conference at Caius College, financed by the National Research Development Corporation (a forerunner of the British Technology Group), which discussed fibre reinforcement and we at Cambridge learnt, through our colleagues from Oxford, of work on directional solidification of eutectics, both in the USA and at Oxford. Work started at Cambridge on this topic. I mentioned that the high rates of work hardening of alloys in the overaged condition appeared to support the idea of fibre reinforcement with a metal matrix.

To begin with, then, I asked Bill Tyson (who joined me with an Athlone Fellowship from Canada) to investigate the validity of Cottrell's idea with a metallic matrix as the adhesive or solder. We tried to produce aligned rods of a very hard phase in a metal by cold working and by directional solidification of a eutectic and other means, since whisker crystals were difficult to use experimentally.

This Bill did but without exciting results. What really got us going was a review by Gene Machlin from Columbia (in my opinion one of the greatest unsung teachers of materials science)[4] and, through it, coming across the work of McDanels, Jech and Weeton.[5] This latter gave us an excellent model system to work on. They had introduced aligned very strong tungsten wires into soft ductile copper, noting that strong wires of many refractory metals are insoluble in the noble metals. Machlin had already recognised the importance of the fibre reinforcement of metals.

Tyson went to work with a will and we produced in a relatively short time a set of principles which could be called the colligative mechanical properties of an aligned fibre-reinforced system. We could explain adequately the strength of the system in terms of the strengths of the two components and how it depended upon volume fraction and on the relative elongation of the two components; on the length of the fibres and more important, the ratio of length to diameter. We explored the idea of minimum and critical volume fraction. I remember with surprise the work being described as being elegant, but I thought it was rather straightforward, but necessary to establish the principles. In the course of doing it we made one of the fundamental discoveries needed. We found a new means of dissipating energy. The fibrous composite, of course, broke. Since we were working with metals and were aware of the emerging concepts of fracture mechanics, we were able to recognise immediately the significance of pull-out at a fracture surface.

When a fibrous composite breaks all the fibres in the fracture plane do not break at once. The fracture surface, therefore, had a jagged appearance and the sliding apart of the two faces or the pull-out of the fibres dissipates a lot of energy and provides a quasi-ductile form of behaviour. This behaviour is not confined to metals and, from Tyson's work, sitting on the numerous quasi-confidential committees I have mentioned, I remember saying that this would probably happen with forms of fibrous composites made from all-brittle systems and would provide an energy absorbing mechanism. It was some years later that I felt my work had been justified when, at a major meeting, one of the inventors of carbon fibre, Leslie Phillips, said, 'and it broke, just like Dr Kelly said it would'.

We were later to discover another, new, means of dissipating energy. This may seem all very obvious to people nowadays, but at the time it was something genuinely new. Other people had described types of pull-out but no one recognised its importance as a source of dissipating energy around a crack.

Tyson and I announced our results at a meeting on new materials organised by Bernal, Bawn, Cottrell and Frank for The Royal Society in 1963.[6] I remember AHC rewrote completely his introductory paper, having read mine, so as to include the concept of pull-out and he made an estimate of the work of fracture of a composite failing by such a mechanism.

Bernal, summing up at the end of the conference, said the contributions on fibre reinforcement were all obvious; he had seen it all done during the war! As, indeed, he could claim, if you know the story of 'Pykrete'.[7*] However, Pykrete was more concerned with reinforcement of brittle matrices than with ductile ones and hence with the second mode of dissipating energy which I will describe later. In my own talk I had to recognise that I was working with a model system and I said that conclusions reached from our experiments were 'that very great strength could be attained using ceramic whiskers embedded in a *metal*. It may be possible to build in adequate impact resistance at the expense of strength in a controlled manner'. During the discussion earlier in the talk, where I was talking about precipitation hardening, I mentioned the utility of alumina and of graphite. I mentioned how useful it would be if one could make fibres of the latter material. Willie Watt told me afterwards that that

* 'Pykrete' was ice reinforced with wood fibres. Pyke proposed to make a floating aircraft carrier of it for emplacement in the mid-Atlantic during the war. I became familiar with the details because the idea was mooted again to us at NPL in the early 1970s to assist the exploitation of oil from the North Sea.

was what stimulated him to go away and invent carbon fibres.[8] I had got to know Watt because, as I told you, I had started work on graphite crystals and he, great inventor as he was, had invented pyrolytic graphite, a quasi-single crystal form of the material, and he had given me some to experiment on.

It was the invention of carbon fibres, almost simultaneously at RAE, Osaka and Cleveland, that introduced what is now called 'hype' into the subject and since then, of course, a number of different strong fibres have become commonplace. In the two tables I compare some of the fibres available prior to 1960 and some of the composites available then with those available in 1986.[9]

4. Multiple Fracture

But there is one more discovery which came from our simple experiments and this was made by Cooper. He was, I think, the second of my research students to work on fibre reinforcement. Since we had discovered pull-out, it was his job to professionalise that and explore it as a means of dissipating energy; a task that he did very well. He also demonstrated, most elegantly, that a crack could not extend if faced with an interface which yielded easily in shear — AHC's point above. George was (is?) a superb experimentalist, somewhat difficult to restrain, and he discovered while at Cambridge, though we did not investigate it fully until we had both gone to the NPL, the principle of multiple fracture. The actual experiment performed by George was as follows. He strained in tension a resin, of elongation to failure, 6% at room temperature and 1% at 77 °K, containing silica fibres coated with carbon of strain to failure 4% at both temperatures. At room temperature, then, the strain to failure of the resin exceeded that of the fibre; at 77 °K the reverse was the case. At room temperature the system was quite brittle but at 77 °K a large extension occurred.

The load extension curve is striking. One of the components breaks in a series of parallel cracks while the other remains and bears the load. The high elongation phase *need not* be continuous. Multiple fracture was produced in the Metallurgy Department here but not analysed until George and I were both at the NPL.*

Here, indeed, was a very striking phenomenon; we had produced ductility. You could produce it with the metal matrix; that is common-

* We applied the results to reinforced cement, showing that it could be quite tough. Fibre reinforced cement had again been stimulated by the development of a fibre, namely Cemfil glass.

Table 1. Some Properties of Various Fibrous Reinforcements and Their Composites Available before 1960

Fibre or Composite	*Young's Modulus, E (GPa)*	*Ultimate tensile strength, UTS (MPa)*	*Specific gravity, SG*	*E/SG (GPa)*	*UTS/SG (MPa)*
Flax	103	690	1.50	69	460
Unidirectional flax/phenolic resin ('Gordon–Aerolite')	34	345	1·35	25	256
Hemp (wet)	34	—	1·50	23	—
(air dried)	57	—	1·50	38	—
(dried under tension)	85	—	1·50	57	—
Ramie (wet)	19	—	1·50	13	—
(air dried)	51	758	1·50	34	505
(dried under tension)	90	—	1·50	60	—
Wood fibre	72*	896	1·50	48	597
Kraft paper/phenolic resin	13	138	1·35	10	102
Asbestos (chrysotile)	159	1379	2·60	61	530
Asbestos/phenolic resin ('Durestos) partially aligned fibres	17	138	1·27	13	109
'E' glass (Ca–Al–borosilicate)	69	3447	2·54	27	1357
Unidirectional 'E' glass/epoxy resin	41	1241	2·05	20	605
Aluminium alloy	70	600	2·8	25	214

* Deduced from the properties of laminates made with these fibres.

Table 2. Some Properties of Various Fibrous Reinforcements and Their Composites Available in 1986

Fibre or Composite	*Young's Modulus, E (GPa)*	*Ultimate tensile strength, UTS (MPa)*	*Specific gravity, SG*	*E/SG (GPa)*	*UTS/SG (MPa)*
'S' glass (magnesium silicate)	86	4481	2·49	35	1780
Unidirectional 'S' glass/epoxy resin	52	1793	2·08	25	862
Boron (boron on W substrate)	379	2758	2·69	141	1025
Unidirectional boron/epoxy	269	1345	1·97	137	683
Carbon (high modulus)	379	1724	2·0	189	862
Unidirectional C fibres/epoxy resin	207	1034	1·60	129	646
Carbon (medium modulus)	234	2992	1·71	137	1750
Unidirectional C fibres/epoxy resin, 60% fibre volume	131	1517	1·55	85	979
Carbon (medium modulus) chopped and aligned/epoxy resin, 55% fibre volume (Ex PERME)	110	1110	1·50	73	740
Asbestos (chrysotile)	159	1379	2·60	61	530
Aligned/epoxy resin (Ex PERME)	93	689 (F)	2·0	46	344
Aromatic polyamide ('Kevlar' 49)	117	2758	1·45	81	1902
Unidirectional Kevlar 49/epoxy resin	83	1931	1·35	61	1430
SiC (Nicalon)	500	4000	2·6	192	1538
Unidirectional Pyrex Nicalon	110	800	2·4	46	333

(F) figure from test in flexure.

place, but to produce it in an all-brittle system was magic. We had made a fully-brittle system show the sort of load extension curve which metallurgists were always willing to find acceptable because the material appeared to be ductile. Of course, it had been done before, as Bernal (see above) would have recognised, but we were the first to name it clearly and to analyse it.

So Tyson and Cooper had each separately discovered a totally new method of dissipating energy in a material so that it could resist fracture: pull-out is one; what is now called microcracking or multiple fracture another. Both will work in all-brittle systems though, as I have said, we discovered pull-out first in a metallic matrix system. They are both vital for understanding the materials science of composites and form contributions from the materials scientist to composite science which is otherwise a very engineering discipline.

Pull-out, as I said, represented a new mode of dissipating energy and gave people confidence that all brittle fibre systems would not be fragile. It has been used to produce energy-absorbing structures and the idea developed using helically conformed fibres. It is one of the principal means by which ceramics may be toughened.[12] Multiple fracture, whether of laminates, or of aligned fibre composites, or of planar matted composites, or of oxide layers on a metal, or of protective coatings, is of the utmost importance. Under various other names, like microcracking, or given the initials MCS, its control is vital in all fibre-composite systems.[10] Control of this phenomenon will probably determine whether or not glasses or ceramics are used in gas turbine engines — as AHC recently opined.

Both those two new methods of dissipating energy in or around a crack were discovered; they were not predicted. I was lucky enough to be able to recognise their significance and to partly and imperfectly analyse them. They illustrate to my mind the central dogma that new materials are always discovered experimentally. They are not predicted. The discovery (there was no prediction) of superconducting oxides reinforces that principle.

References

1. J. C. Slater, *Quantum Theory of Matter*. McGraw Hill, New York, 1959.
2. A. H. Cottrell, *Proc. Roy. Inst.* **38**, 346, 1960.
3. C. P. Talley, *J. Appl. Phys.* **30**, 1114, 1959.
4. E. S. Machlin, Status Report on Non-Metallic Fibrous Reinforced Metal Composites, Materials Research Corporation Contract NOW 61–0209–c, 1961.
5. D. L. McDanels, R. W. Jech and J. W. Weeton, *Metal Progress*, **78**, 118, Dec. 1960.

6. J. D. Bernal, C. E. H. Bawn, A. H. Cottrell and F. C. Frank, 'A Discussion on New Materials', in *Proc. Roy. Soc.* **A282**, 1–156, 1964. The meeting took place on 6–7 June, 1963.
7. See M. F. Perutz, *Discovery*, **9**, 147, 1948, for the story of Pykrete; and C. E. Phillips and R. C. A. Thurston, 'Some Mechanical Properties of Pykrete at Temperatures below 0 °C', NPL Report H/2509, Sept. 1943, for some of the data.
8. W. Watt, L. N. Phillips and W. Johnson, 1964, British Patent No 1 110 791.
9. A. Kelly, *Proc. Roy. Inst.* **136**, 794, 1986; see also Ref. 11.
10. A. Kelly and N. H. Macmillan, *Strong Solids* (3rd Edn), O.U.P., 1986.
11. P. McMullen, *Composites* **15**, 222, 1984.
12. A. G. Evans, 'High Toughness Ceramics and Ceramic Composites', pp. 51–91, in 10th Risø International Symposium on Metallurgy and Materials Science (Eds J. B. Bilde Sorensen *et al.*), 1989.

11 Strengths of Grain Boundaries in Pure Metals

A. H. Cottrell

Research Associate, Department of Materials Science and Metallurgy, University of Cambridge

Abstract

The relative surface energies for brittle fracture along grain boundaries or along crystal planes, at low temperatures, are estimated and used in a criterion for the relative strengths of boundaries and cleavages. It is concluded that the boundary is weaker than the crystal in a wide range of metals; that it becomes weaker as the ratio of shear modulus to bulk modulus increases; and that brittle pure metals, such as iridium and molybdenum, fracture preferentially on grain boundaries. The critical modulus ratio is in all cases lower than that for the ductile–brittle cleavage transition.

Introduction

It is commonly assumed that arbitrary high angle grain boundaries in metals are strong at low temperatures, unless weakened by segregated impurities. This assumption of intrinsic strength is surprising, however, in the light of the atomic irregularity of such a boundary, across which many atoms are finitely separated from their opposite neighbours. Nor is there support for it from the soft metals, since the low yield strengths of these usually prevent the applied stress from achieving levels at which tensile strength of a boundary might be put to the test.

In the past, purported examples of intercrystalline fractures in nominally pure metals, at low temperatures, have usually been dismissed on the grounds that certain impurities, e.g. sulphur in iron, in amounts too small for detection by conventional analysis, are known to segregate to grain boundaries and thereby weaken them. However, the application of Auger spectroscopy has changed this and examples have now been reported of brittle intercrystalline fractures in which the exposed faces showed no impurities when examined using that method; e.g. iridium,[1,2] rhodium,[1] molybdenum,[3] Ni_3Al,[4] Fe–Mn alloy.[5] By contrast, high purity iron has shown no intercrystalline fractures even at the lowest temperatures.[6]

The standard theory of intercrystalline fracture is based on the comparison of surface and grain boundary energies. The complexity of atomic structure in an arbitrary high angle boundary precludes accurate atomistic calculations of the ideal tensile strength. To bypass this problem, Hondros and McLean[7] assumed that, in the first approximation, this strength is proportional to the surface energy. This assumption is reasonable provided that there are no large variations in the shape of the atomic force-displacement function for the various modes of failure being compared, a condition which is expected to be satisfied in the present case, i.e. brittle grain boundary fracture versus brittle crystal cleavage. Thus, of interest is the ratio of the surface energies for grain boundary fracture and crystal cleavage given by

$$R = (2\gamma_{sb} - \gamma_b)/2\gamma_{sc} \tag{1}$$

where γ_{sb} and γ_{sc} are the values of free energy (per unit area) of one face of the surface exposed by a boundary fracture and by cleavage, respectively, and γ_b is the free energy of the boundary. When $R > 1$, cleavage would be expected rather than boundary fracture and vice versa when $R < 1$. With typical values of surface and boundary energies it is obviously possible that boundaries in at least some metals are intrinsically weak. Hondros and McLean[7] recognised this, but referred to other effects, such as plastic flow at crack tips, to maintain the view that boundaries in pure metals are strong.

Comparison of boundary and cleavage energies

The anisotropy of surface energy, i.e. $\gamma_{sb} \neq \gamma_{sc}$, must also be taken into account. Since a cleavage crack will obviously choose the crystal plane of lowest surface energy, at least in crystals of high symmetry, it is reasonable to take γ_{sc} as the minimum free surface energy γ_s^{min}, in which case $\gamma_{sb} \geqslant \gamma_{sc}$. For an arbitrary high angle grain boundary, which exposes atomically rough faces when fractured, it also seems reasonable to take γ_{sb} as the maximum free surface energy γ_s^{max}. Experimental work, e.g. by McLean,[8] shows that at the melting point λ ($=\gamma_s^{max}/\gamma_s^{min}$) ranges from about 1·015 (Cu) to 1·05 (Au, Pt) for various metals examined (Ag, Au, Co, Cu, γ-Fe, Ni, Pt); and that it increases at lower temperatures. Extrapolation to room temperature leads to $\lambda \simeq 1{\cdot}2$, which is used here when examining the ratio

$$R = \lambda - \frac{\gamma_b}{2\gamma_{sc}} \tag{2}$$

This ratio admits the possibility, when $\gamma_b < 0{\cdot}4\gamma_{sc}$, that grain boundaries are intrinsically stronger than cleavage planes.

Now, relative values of γ_b and γ_{sc} are required. While γ_s is known fairly well for many metals, good data for γ_b are scarce. Therefore, to proceed, γ_b and γ_{sc} are here related to elastic constants. For γ_b, the standard dislocation theory of tilt boundaries is considered, which, although rigorous only for low angle boundaries, is also well known to give fairly good values for high angle boundaries[9]

$$\gamma_b = \frac{\mu b}{4\pi(1 - v)}\,\theta_m \tag{3}$$

where θ_m is a characteristic tilt angle ($\simeq 0{\cdot}5$ rad), μ is the shear modulus, v is Poisson's ratio, and b is the unit slip distance.

The only feature of equation (3) to be used below is that $\gamma_b \propto \mu$. It can be seen from equation (3) that the main contribution to γ_b is from the resistance to shear and that the resistance to dilatation, as measured by the bulk modulus K, plays only a concealed minor role through the factor $(1 - v)$. By contrast, the main contribution to γ_s is expected to be from K, not μ. This is because the process of producing free surface by fracture involves the passage of a crack; and the elastic field at the crack tip is one of biaxial tension and plane strain,[10] the elastic resistance to which is not much different from that to pure dilatation. Thus, R is expected to decrease as μ/K increases and a transition from cleavage to boundary fracture is expected to occur when μ/K exceeds a critical value.

Values of μ/K, from single crystal ratios, compiled by Hecker *et al.*[1] are given in Table 1. Here, μ is the average shear modulus on the slip planes, (111) for fcc and (110) for bcc. The ratio μ/K has of course frequently been used as a criterion for the ductile–brittle transition, originally reported by Pugh.[11] In the present work, a second transition, from transcrystalline to intercrystalline weakness, at some value of μ/K, is proposed.

To establish this value, from equation (2), μ/K is first related to γ_b/γ_{sc} for a particular metal, then this is used to order the other metals in proportion to their value of μ/K. The calculations of Chen *et al.*,[12] using a high quality interatomic potential, of symmetrical tilt [001] boundaries, gave an average value of $\gamma_b = 1210\,\mathrm{mJ\,m^{-2}}$ for nickel. From their corresponding calculations of the Griffith cohesive energy, $2\gamma_s - \gamma_b$, the average value of $2\gamma_s = 4100$ can be deduced. Assuming, from $\lambda = 1{\cdot}2$, an anisotropy spread of 10% about this average, $2\gamma_{sc}$ can be taken as 3700. It then follows from Table 1 that

$$\alpha = \frac{\gamma_b}{2\gamma_{sc}(\mu/K)} \tag{4}$$

Table 1. Values of parameters obtained in present work

Parameter	Au	Ag	Cu	Pt	Ni	Rh	Ir	Nb	Ta	V	Fe	Mo	W	Cr
μ/K	0·114	0·192	0·223	0·237	0·343	0·518	0·520	0·246	0·306	0·322	0·331	0·476	0·591	0·819
R (equation (5))	1·09	1·02	0·99	0·97	0·87	0·71	0·70	0·97	0·91	0·89	0·88	0·75	0·71	0·42
$(\mu/K)_{DBT(C)}$	0·36	0·43	0·57	0·38	0·49	0·39	0·32	0·59	0·55	0·65	0·56	0·35	0·45	0·68
$(\mu/K)_{DBT(B)}$	0·39	0·44	0·56	0·37	0·43	0·28	0·22	0·57	0·50	0·58	0·49	0·26	0·32	0·29

DBT(C) and DBT(B) are ductile–brittle transitions for cleavage fracture and boundary fracture, respectively.

yields $\alpha = 0{\cdot}95$. This value is used in

$$R = \gamma - \alpha(\mu/K) \tag{5}$$

to deduce the values of R given in Table 1. It can be seen that, for all the metals considered, except gold and silver, the grain boundaries are expected to be weaker than the grains, although in all cases except Rh, Ir, Mo, W, and Cr, the difference is small and probably within the uncertainties of the calculation.

Putting $R = 1$, the condition for the cleavage–boundary transition (CBT) is that

$$\left(\frac{\mu}{K}\right)_{CBT} = \frac{\gamma - 1}{\alpha} \tag{6}$$

i.e.

$$\left(\frac{\mu}{K}\right)_{CBT} = 0{\cdot}21$$

when the above values of λ and α are used; this is an unusually low value of μ/K. This condition can be compared with that for the ductile–brittle transition (DBT). Rice and Thomson,[13] building on the earlier work of Pugh[11] and Kelly *et al.*,[10] expressed the ductile–brittle transition in terms of the difficulty of blunting a crack tip by the emission of dislocations. This led them to the criterion

$$\frac{\mu b}{\gamma_{sc}} \gtrsim 10 \tag{7}$$

for brittleness. This can be rewritten in the form

$$\left(\frac{\mu}{K}\right)_{DBT(C)} = \frac{10\gamma_{sc}}{bK} \tag{8}$$

as the criterion for the ductile–brittle transition in the case of cleavage (C) fracture. There is a corresponding relationship

$$\frac{\mu}{K}_{DBT(B)} = R\left(\frac{\mu}{K}\right)_{DBT(C)} \tag{9}$$

for the ductile–brittle transition in the case of boundary (B) fracture. Values obtained using the data compiled by Hecker *et al.*[1] and taking $\gamma_{sc} = 0{\cdot}9\gamma_s$ are given in Table 1.

Discussion

It can be seen that, in every case, the transition from ductile to brittle cleavage is expected to occur at a higher value of μ/K than the transition from cleavage to boundary weakness. In other words, cleavage would not be expected in a polycrystalline pure metal: at low temperatures the metals considered are either ductile (Au, Ag, Cu, Pt, Ni, Nb, Ta, V, Fe) or are brittle due to intercrystalline failure (Rh, Ir, Mo, W, Cr).

The large difference in the two ductile–brittle transition values for metals of high μ/K ratio suggests that, in principle, there could be examples where the polycrystal is brittle, due to grain boundary weakness, whereas the single crystal is ductile, although no such examples are predicted in Table 1. However, it is known that polycrystalline Ni_3Al is brittle due to grain boundary failure (in the absence of boron doping), whereas the single crystal is very ductile.[4, 12] Also, single crystals of iridium become brittle only at much lower temperatures than polycrystals.[1]

Iron is an interesting example. Table 1 predicts that it should be (marginally) ductile, which seems to be so in the superpure metal.[6] Presumably, the prevalence of brittle cleavage fracture in the commercial metal at low temperatures is due to two effects: hardening by impurities, which reduces dislocation activity in the vicinity of crack tips and so takes the metal above the ductile–brittle transition; and strengthening of the grain boundaries by segregated carbon, which takes the metal below the cleavage–boundary transition.

Of course, there are considerable uncertainties in the values of λ and α in equation (5), from which the values of R given in Table 1 are deduced. The rather large difference between the values of R for the brittle metals Rh, Ir, Mo, W, and Cr and for the others in Table 1 does, however, suggest that the conclusion — that the polycrystals of these pure metals (and only these) undergo brittle grain boundary fracture at low temperatures — may be fairly definite, despite these uncertainties.

Acknowledgments

It is a pleasure to thank Professor D. Hull and his colleagues, especially Dr J. A. Charles and Mr G. C. Smith, in the Department of Materials Science and Metallurgy, University of Cambridge, for their friendly support during the course of this work. I am also grateful to Professor V. Heine, University of Cambridge, for bringing the problems of grain boundary fracture to my attention and to Dr R. W. Cahn also at Cambridge for a useful discussion.

References

1. S. S. Hecker, D. L. Rohr and D. F. Stein, *Metall. Trans.* **9A**, 481, 1978.
2. C. L. White, R. E. Clausing and L. Heatherly, *Metall. Trans.* **10A**, 683, 1979.
3. S. Suzuki, H. Matsui and H. Kimura, *Mater. Sci. Eng.* **47**, 209, 1981.
4. C. T. Liu and J. O. Stiegler, *Science* **26**, 636, 1984.
5. H. J. Lee and J. W. Morris, *Metall. Trans.* **14A**, 913, 1983.
6. H. Matsui, S. Moriya, S. Takaki and H. Kimura, *Trans. Jpn Inst. Met.* **19**, 163, 1978.
7. E. D. Hondros and D. McLean, in 'Grain boundary structure and properties', (ed. G. A. Chadwick and D. A. Smith), chap. 10; Academic Press, London, 1976.
8. M. McLean, *Acta Metall.* **19**, 387, 1971.
9. F. R. N. Nabarro, 'Theory of crystal dislocations', Oxford University Press, Oxford, 1967.
10. A. Kelly, W. R. Tyson and A. H. Cottrell, *Philos. Mag.* **15**, 567, 1967.
11. S. F. Pugh, *Philos. Mag.* **45**, 823, 1954.
12. S. P. Chen, A. F. Voter and D. J. Srolovitz, in MRS Symposium 'High-temperature ordered intermetallic alloys II' Vol. 81, (ed. N. S. Stoloff *et al.*), 45; Materials Research Society, Pittsburgh, PA, 1987.
13. J. Rice and R. Thomson, *Philos. Mag.* **29**, 73, 1974.

Discussion

PAPER 2: Theory of the Yield Point and of Strain-Ageing in Steel

Professor D. V. Wilson: Well, it's certainly fascinating to look back on a few of the many lines of enquiry which started from Alan Cottrell's elucidations of dislocation — solute interactions. However, just at the moment, I would also like to thank Norman Petch. He has informed us, guided us and sometimes provoked us into activity over a period of more than 40 years. His contributions to understanding the macroscopic behaviour of polycrystalline metals are unique. They are also perennial. Those of us who passed the milestone of 70 years a few years ago may be forgiven for being a little envious. Has Norman Petch also discovered the secret of eternal youth? Be that as it may, thank you Professor Petch: I don't know what we would do without you.

Professor W. D. Biggs: I would just like to go back to comment that was made by Bruce Bilby, and that was about Alan's ability to make a suggestion which set you thinking and then you produced an experiment and to your astonishment you found that the suggestion was right.

I well remember listening to Sir Alan giving us a colloquium in which, as was his wont to do, he developed a new theory on the board as he went along — which made life somewhat confusing for new research students who, afterwards, couldn't remember what the theory was all about. But this one happened to be about order–disorder transformations, and the fact that he thought it highly possible, that if you had an ordered structure in such a form that when you produced slip you had to push two similar atoms across each other, the yield point would be higher than when you had to push dissimilar atoms across each other. And Trevor Broom and I were talking about this at coffee. I happened to have a machine that could detect a yield point in anything. (That, incidentally, is a separate story.) But Trevor and I talked about this. I had the machine. Trevor had some copper-gold wires and he had a resistivity apparatus to measure the degree of ordering, and within 3 days we had, in fact, discovered that the yield point increased when you had this particular situation. Sir Alan had by then disappeared to America to lecture, so we sent him the information as fast as possible and he managed to put it in his lecture. We published it, I think, in the *Phil. Trans. of the Royal Soc.* and it was the best piece of research I have ever been involved with! From start to finish, including publication, it took four weeks; and that I think was a record, and all due to a chance remark from Sir Alan.

PAPER 3: Ductile and Brittle Failure in Steel/Pressure Vessel Assessments

Dr J. F. Knott: Thank you, Mike, for your paper, which I support completely, bar a few minor details. It is 30 years since I came to Cambridge as a research student to work under Sir Alan's supervision, and I would like to pay tribute to his inspiration throughout so much of my research. The reason that I came to Cambridge from Sheffield is that I was urged to do so by Trevor Churchman, from CEGB at the time. I well remember the day when the letter came, offering me a 'burst rig' to come to Cambridge to do research. It turned out later to be a bursary, for which I was most grateful. I had a letter from Sir Alan during the Long Vac. saying that what I would be doing would be to try to look for the evidence of dislocations producing a crack nucleus, using the transmission electron microscope, and so I read up all I could about that. When I arrived I was to be given material on macroscopic plasticity by Green and Hundy, and the draft paper with Professor Bilby on continuous dislocation distributions and crack opening displacements. That changed the whole approach. I had to do a lot more reading and then moved into brittle fracture and the various topics referred to in Professor Burdekin's paper. I would like to pay tribute here to my research students, who have done absolutely excellent work in the area.

There are two comments I would make. One is that I think what we have been able to do is to be quantitative about brittle fracture and microstructure, and that has been tremendous from the materials point of view because we have been able to tailor our materials to get better microstructures. This follows Professor Petch's line: 'look at what the material is doing and try and improve that'. The second comment, and I think it is witnessed by the presence of Professor Burdekin here, is that we have been able to get good links between materials and structural engineering: we have good rapport, we talk the same language. That is something that has not yet been achieved so well in other parts of the world. Through all this and through the later exercises of the PWR inspection validation, and, I may say, with respect to new materials developments for Rolls-Royce, Sir Alan's guiding influence and inspiration has been with us all. Thank you.

Dr R. E. Dolby: As usual, Mike Burdekin has given us a presentation of great clarity. The one thing he failed to clarify was his own substantial role in developing the UK into the number one position in the world for engineering assessments of defects in welded structures. In fact, I believe the UK has been in the world lead in this technology area for many years. Not only did the UK develop one widely accepted methodology, we developed two! — the PD 6493 and R6 approaches. It is very gratifying to see them now coming together into a single document.

We should congratulate Mike for, in a recent International Institute of Welding Conference in Helsinki, he pushed through this unified UK document on the engineering significance of defects where at least 30 nations were represented from the developed world. That was a very significant step forward which should lead to a most important international standard before long — a document which has, as one of its origins, Alan Cottrell's and Alan Wells's early ideas on crack tip opening displacement from over 30 years ago.

Everything goes in cycles, and one thing that has struck me recently is that whilst these cycles of rediscovery are normally 7 to 10 years, it has taken 30 years of fracture research to get back to the original local fracture approach of Alan Cottrell and John Knott. I am not sure why it has taken that long, but the 'local' approach is certainly fashionable in France at the moment. Is it because the French researchers did not read the literature from the early sixties, or is there something really new?

Professor F. M. Burdekin: I think the big difference is the development of finite element techniques, which enabled people to believe anyway that they have more accurate information about stress fields locally; particularly in the elastic–plastic situation and in the three-dimensional situation, although much of the local approach work is actually 2D at the moment. It does offer enormous potential in that way. Like you, I think as we get older we tend to see these cycles coming round: things being re-invented. They do get improved each time and given a little kick to maintain the momentum as they go.

Dr B. L. Eyre: Professor Burdekin referred to the crucial role played by Sir Alan Cottrell in stimulating the very careful examination carried out in this country of the structural integrity of PWR pressure vessels. The work carried out by Walter Marshall and later Peter Hirsch in addressing the concerns voiced by Sir Alan resulted in a greatly improved understanding of the factors influencing the toughness of the A533B and A508 steels used for the pressure vessels. It also focused on the very stringent requirements of quality assurance and led to a considerable refinement of the inspection techniques and to the concept and application of inspection validation. It is clear that the UK nuclear industry, particularly CEGB and UKAEA have led the way and the approaches adopted will undoubtedly be of benefit to other plant where structural integrity and safety are of prime concern.

There is also a growing recognition of the need to incorporate a greater degree of passive safety in next generation nuclear plant — particularly pressurised water reactors — and to rely less on engineered safety provisions. I believe that the debate stimulated by Sir Alan on pressure vessel integrity under reactor transient conditions was one important factor in promoting this very positive development.

PAPER 4: Dislocation Reactions

Dr L. M. Brown: Well, I first wanted to say that in 1956, as a student at the University of Toronto doing engineering physics, I read Alan Cottrell's books: *Theoretical Structural Metallurgy* and *Dislocations and Plastic Flow in Crystals*. And, certainly, a large part of the reason for wanting to come to England was to study with him. I came to Birmingham but by then he had left. I think what was so inspiring about those books was the promise they held out — that you could do a 'back of envelope' calculation and arrive at, say, the work-hardening rate of a polycrystalline metal, or to simplify the problem still further, the stage II hardening of a single crystal of a typical ductile metal. It was that promise that was so exciting, and 32 years later it still seems to me a most interesting proposition. What I want to ask and would very much like to hear discussed is: considering all we now know about the fundamental mechanisms of plasticity, so elegantly demonstrated by Dr Whelan, can we make such a back of envelope calculation, and if we can't, why is it that we can't do it at this point?

Dr M. J. Whelan: I am always very much in favour of back of envelope calculations, not having used a computer myself for many years. But I would agree with everything you say: that Alan Cottrell's book has proved to be such a great inspiration to generations of students who have gazed into it. I have been producing a book, *Worked Examples in Dislocations*, and I had to consult Sir Alan's book on detail frequently over the last year or so. I still find it very useful to look at.

Professor C. J. Humphreys: Let me go back to a conference 20 years ago, when I think Sir Alan must have been a mere 50! Peter Hirsch had delivered a conference paper by Whelan and Hirsch on electron microscope observations of dislocations, in a somewhat similar vein to the presentation we have just had from Mike Whelan, when someone from the audience stood up and asked him: 'What use is all this knowledge of dislocations?' It was the sort of deceptively innocent question that, posed 20 years ago, would have reduced many mortals to silence. However, after a moment's thought, Peter replied that he would not give any specific examples but what was really important was that a knowledge of dislocations had fundamentally changed the way we think about materials.

That was a very wise answer; however, the point I would like to make is that because of research in the last 20 years, based on earlier work of Alan Cottrell, we can now go much further than the statement of Peter Hirsch and 'engineer' dislocations to give materials specific properties. Let me give a very elegant example of dislocation engineering that this year appeared in a commercial semiconductor device. The work is based on a seminal paper of Alan Cottrell written 40 years ago on the attraction of solute atoms to dislocations, and Norman Petch in his talk referred earlier to this paper. For certain semiconductor devices you need the surface layers of silicon (in which the devices are made) to be ultrapure, much purer than can be achieved by zone refining. How do you get the surface that pure? The answer is to introduce a controlled array of dislocations into the silicon sufficiently close to the surface that they attract the impurity atoms and 'getter' the surface clean, but sufficiently far from the surface that they do not impair device performance. This can now be done using strained epitaxial layers to introduce a controlled array of misfit dislocations at a given depth. However it has taken 40 years from the seminal work of Sir Alan in this field to the first commercial device utilising dislocation engineering for gettering impurities. The emerging field of dislocation engineering is based very much on the early work of Alan Cottrell which is quite breathtaking for its vision and breadth.

PAPER 5: The Nature of Grain Boundaries in Ordered Alloys

Professor R. Maddin: I am most appreciative for the opportunity given me by the organisers of this memorable symposium, Gerry Smith and Jim Charles, to make a few remarks. It is appropriate to follow Professor Cahn, since the remarks refer back to the time I exchanged positions with Professor Cahn during 1954 at the University of Birmingham. My association with Sir Alan began at that time and our relationship was soon cemented when little Geoffrey Cottrell hit my daughter on the head with a toy.

I studied metallurgy at a University where perhaps the only use of the term 'dislocation' referred to an American football injury; it therefore was with some trepidation that I sat in on his lectures on dislocations given for research students. The clarity and articulation of his lectures were immediately apparent. Sir Alan and I began our collaboration with an experiment that led to a publication of a paper a year later in *Phil. Mag*. We noticed that quenching small (*ca*. 3 mm) Al single crystals from close to the melting point increased the yield strength by at least a factor of ten. Not being convinced of our first observations, Sir Alan and I designed an experiment to increase both the quenching rate and the range; we were to enclose a square, 3 mm on an edge, single crystal of Al in a pure alundum sheath the open end of which contained a thin wire of nichrome. The Al crystal also contained a wire so attached that at the quenching the crystal could be extracted from its sheath very quickly so that liquid nitrogen could be poured in large quantity over the crystal. The experiment was a disaster and to this day neither Sir Alan or I are sure which of us messed up. On my side of the Atlantic, it was Sir Alan who caused the sheath to be quenched by mistake but I have learned that on this side of the Atlantic, it was I who cut the wrong wire. Fortunately, however, our observations were confirmed by other experiments and other researchers.

Many thanks for the opportunity to be here and take part in these historic proceedings.

Professor D. M. R. Taplin: This paper on grain boundaries and ordering leads one back once again to the largely discredited ideas of the Australian, Walter Rosenhain on amorphous cement grain boundaries and creep. As Walter Boas (also of Australia) perhaps most cogently argued, the whole structural basis of the amorphous cement theory was flawed — yet this vintage concept provided an overview and synthesis which still has some merit today. We should be careful not to discard such ideas too readily, because on the one hand, the creative thought processes involved are often instructive and on the other later ideas sometimes bring back to the leading edge much older ideas in modified form. The work of Sir Alan has sometimes been found by later experiments not to be perhaps as universal as was originally envisaged, but the creative genius of the ideas is such that they have an enduring quality. My question is, could Professor Cahn kindly comment on the thinking behind the amorphous cement theory as it applies to the work discussed in his paper?

Professor R. W. Cahn: The status of Walter Rosenhain's complex of hypotheses concerning amorphous cement at grain boundaries and along slip planes is best compared to the status of dislocations in their very early days, before Sir Alan Cottrell came to grips with them. Before the geometry and elastic interactions of dislocations had been properly examined, physical metallurgists who read about these intriguing new entities (the very existence of which had not yet been verified) pounced on them with glee and they were invoked to explain all sorts of phenomena. The trouble was, the proposed explanations were vague and wholly qualitative and they did not incorporate that most essential feature of a scientific theory (as Sir Karl Popper has taught us) — the quality of falsifiability. When a theory ('notion' would be a better term) is put forward in such a form that, even in principle, no experiment or other theoretical analysis is conceivable the result of which might disprove it, it is not a proper theory and has no explanatory power. This, unfortunately, is what Rosenhain's amorphous cement notions were like. Amorphous cement was supposed to be disorderly and so were the debates which followed on the presentation of Rosenhain's lectures: to each objection there was a further hypothesis, until notion was piled on notion to form an ever more tottery edifice. I have noticed that the proponents of notions which fall foul of Popper's test quite often reveal a striking characteristic: when the notions are challenged, the proponents are apt to lose their temper, and this is clearly detectable when one reads Rosenhain's furious response to critics, or W. A. Wood's acerbic response to critics of his papers in the 1950s in which he denied the existence of dislocations!

As I remarked in the spoken form of my paper, the quintessence of Sir Alan's approach to theorising is to be found in his theory, with Bruce Bilby, to interpret discontinuous yield and strain-ageing in mild steel. True, the subtleties of upper and lower yield points and the propagation of Lüders bands continued to be refined for some years after Cottrell and Bilby wrote their key papers but the essential idea, the elastic interaction between an edge dislocation and interstitial solute atoms clustering around it, was never seriously challenged (though I suspect Rosenhain would have challenged it had he still been alive!). What has long impressed me as the quintessence of scientific method, however, was the strictly quantitative derivation of the predicted kinetics of strain-ageing on the basis of this precisely defined model: the kinetics could then be examined experimentally by measuring (using a Snoek pendulum, itself dependent on sophisticated theory) the rate at which carbon atoms ceased to be free agents and became captive satellites of dislocations. If this measurement had revealed a time dependence clearly in disagreement with the Cottrell/ Bilby theory, that would have been the end of it. That is what falsifiability means.

Reverting now to the particular terms of David Taplin's question, I really don't believe that the amorphous cement notion is relevant to the current debate as to whether local segregation to grain boundaries in an ordered alloy may disorder the adjacent lattice under appropriate conditions. 'Disordering' here refers strictly to the loss of a superlattice — there is no suggestion that crystalline order is lost and replaced by an amorphous disposition of atoms! There is a sound body of theory relating the balance between atomic order and disorder to composition and temperature, and a separate body of theory and experiment governing the tendency of different kinds of atoms to segregate to grain boundaries, so there is a reasonable prospect that, if localised disordering near grain boundaries in intermetallic compounds is eventually convincingly confirmed, it can be interpreted in terms of a proper falsifiable model.

PAPER 6: Radiation Damage and Point Defects

Dr A. T. Churchman: I am a bit surprised at some of the omissions from Ron Bullough's reprise of Sir Alan's time at AERE. There are two beautiful experiments which Ron did not mention, one of which proved critically important to the embryo civil nuclear power programme — UK's response to 'Atoms for Peace'. But I suppose they did occur before Ron's day.

The first experiment demonstrated that irradiation-induced point defects were attracted to dislocations, an extension of the Snoek damping which Robert Cahn mentioned. Sir Alan suggested that Robert Barnes should irradiate some annealed copper rods, and, if the conditions were appropriate, Robert would find that the copper would ring as if it were a bronze bell. Sure enough the point defects locked the dislocations, thus removing their ability to damp out imposed vibrations.

The second, and of much more import, gives the answer to Colin Humphries, and demonstrates that one can, from such studies, devise simple critical experiments from which one may predict future behaviour of engineering structures. On the list of papers which Bruce Bilby showed to you (but did not mention), was one by Cottrell and Roberts. Here Sir Alan was following up the well established finding that for Uranium, thermal cycling through the phase change or irradiation produced length changes: that this phenomena was suppressed in Pippa fuel bars by creating a random fine grain size. Sir Alan argued that, at least under irradiation this would result in a build-up of internal stress until they equalled the yield point of the Uranium. The Cottrell/Roberts U spring experiment elegantly demonstrated the consequent irradiation-induced creep. He knew that the Pippa fuel elements in the newly commissioned Calder Hall Magnox reactor were long thin bars of U, stacked one on top of the other, and were thus prone to strut bowing if there was any plastic relaxation in the U. Sir Alan predicted that if the Pippa Mk1 fuel elements were left in the reactor past 1000 MWd/t, half their intended dwell time, they would bow sufficiently to jam themselves in the graphite channels: if so this would be to say the least an unfortunate predicament for the operators to find themselves in. Fortunately the relevant managing director was an engineer, no less than Christopher Hinton as he then was, who, although a terror, was wont to believe in good experiments. Against the operator's better judgement, Hinton ordered a shut down before 900 MWd/t was reached. Sure enough, the key elements were found to be starting to bow. The replacement Pippa Mk1 was braced with longitudinal Magnox struts which Sir Alan proposed as the cure: Calder Hall still generates successfully.

I maintain that these were two beautiful experiments; the second demonstrating how powerful a critical scientific experiment can be to commercial success, if one is perceptive enough to predict its consequences in real conditions.

Dr B. L. Eyre: Dr Bullough has referred to the pioneering work carried out at Harwell by Sir Alan and his colleagues on irradiation hardening and embrittlement in pressure vessel steels in the late 1950s. At about the same time I joined CEGB as a member of the nucleus team for Berkeley Nuclear Laboratories and I started research into irradiation damage and the effects on mechanical properties of iron. I was in fact extremely fortunate in meeting Dr Constance Tipper at an early stage in this work. She was just about to retire from her Readership in the Engineering Department here at Cambridge and she gave me a number of single crystals of pure iron she had prepared. I irradiated these in the Herald Reactor at Aldermaston and carried out tensile tests and transmission electron microscopy on the iron single crystals. Optical metallography of the tensile specimen surfaces showed the slip bands to change from diffuse wavey traces in the unirradiated specimens to sharp straight traces for the irradiated case.[1] I remember very well showing these results to Sir Alan, who was a consultant to CEGB at the time, and he suggested that the change in slip morphology was a consequence of dislocation channelling. He encouraged me to look for more direct evidence. Subsequent transmission electron microscopy carried out with the help of Kingsley Williamson in fact provided evidence for point defect clusters which interacted with and were swept up by mobile dislocation in the thin foils.[2] This was the first direct observation of channelling in irradiated iron, which was later confirmed more convincingly on iron irradiated to higher doses.[3]

On the actual mechanism of hardening, I believe the important point defect clusters to be interstitial loops in neutron irradiated iron. The available evidence suggests that the vacancy rich cascade centres do not collapse to form vacancy loops. On the other hand, the recent very beautiful electron microscopy evidence referred to by Dr Bullough shows very clearly that solute atoms, certainly Cu, but also possibly other impurities such as P, play a key role in enhancing hardening by forming coherent clusters. The irradiation-induced vacancies may play an important role in promoting irradiation enhanced diffusion of the solutes to form solute clusters, possibly in association with interstitial clusters. Hence, the observed sensitivity to dose rate which is consistent with heterogeneous nucleation.

References

1. B. L. Eyre, *Phil. Mag.*, 1962a.
2. B. L. Eyre, *Phil. Mag.*, 1962b.
3. B. L. Eyre and A. F. Bartlett, *Phil. Mag.*, 1965.

PAPER 7: 'The Metallic State — Revisited'

Dr J. E. Evetts: Peter has shown us how the concept of metallic behaviour has broadened and deepened over the years. There are now a number of oxides and nitrides that are electronic conductors; they have a low electron density but are clearly metallic. I hesitate, however, to call them metals, and prefer perhaps to use the term semi-metal to indicate something closer to the semiconducting state. In 1986 we learnt that metallic oxides can also be superconducting with transition temperatures over 100 K; this discovery alerted the world to the potential for applied superconductivity in technology. Sir Alan had made this connection some twenty-five years earlier! In 1961 Professor Cottrell, as Head of the Metallurgy Department at Cambridge, set up the first Applied Superconductivity Group outside a low temperature physics environment. Within two or three years the first applications of superconductivity began and twenty years on we have a major UK industry. Sir Alan's initiative helped to lay the basis for this industry, and more parochially set in motion the train of events that enabled Cambridge University to bid successfully for the Interdisciplinary Research Centre in Superconductivity. To return to the oxide superconductors, it is possible that the low electron density they exhibit may be able to be put to use in a new class of field effect devices. Very thin films of these oxides can already be deposited; if such films can be patterned lithographically on a fine enough scale an applied electric field may be able to switch the superconductivity on or off. Such a device would operate right on the borderline of the metallic state embodying at the same time aspects of the superconductor, the semi-metal and the semiconductor.

Dr P. P. Edwards: Yes, Jan, you have brought out a very important point. Surely, no-one in their wildest dreams would have believed that this black, dense ceramic oxide could be a 90 K superconductor. Put it in a vacuum furnace at several hundred degrees for a quarter-of-an-hour (or so) and it transforms to a semiconductor. So you are quite correct, it is a question of how to chemically manipulate oxides, and other materials, across the transition region separating metallic *vs* non-metallic states. In other words, reduce the electron density so as to move the material closer and closer to the Metal-to-Non-metal Transition. High temperature superconductivity may be one of the most exotic features of this transition.

PAPER 8: Microstructural Characterisation of Defects and Mechanisms

Professor D. G. Brandon: In the summer of 1958 the International Electron Microscope Congress was held in Berlin. Erwin Muller, who had been recruited for American science after the war, returned to his home ground and lectured at the Congress on Field Ion Microscopy. His atomic resolution images of refractory metal surfaces brought him an ovation and a request to repeat his lecture a second time. In 1958 Alan Cottrell was appointed Goldsmiths' Professor of Metallurgy at Cambridge. I had been a research student in the Department for two years. The thin film revolution in transmission electron microscopy (TEM) was in full swing and our enthusiasm for tackling every conceivable metallurgical problem by TEM left little motivation for developing alternative techniques. Nevertheless, a few months later my research supervisor, Jack Nutting, called me into his office and asked if I would be interested in taking charge of a research group under Alan Cottrell once my thesis was completed. The aim would be to study radiation damage at atomic resolution and the primary tool of investigation would be the field ion microscope (FIM). The funding would come primarily from the United Kingdom Atomic Energy Authority.

In retrospect, I find it little less than incredible that, from an initial position of total ignorance and with totally unrealistic expectations, field ion microscopy nevertheless grew and flourished in Cambridge. The autumn of 1959 found me installed in a laboratory which had originally been part of (I think) Colloid Science. I was assigned a technician by the name of Joe Reich, who stayed with the group until his retirement, long after I had left, and whose life story was to contribute to the education of all of us.

The first research students were Mike Southon, Piers Bowden and Mike Wald. Mike Southon was a physicist whose insights into the mechanisms of field ionisation and field evaporation were to prove invaluable. Piers Bowden was the odd man out. Before starting to build our first field-ion microscope we had decided to hedge our bets and initiate some work on the generation of point defects by low energy ion bombardment, using TEM as the investigative tool, in order to have some point of contact with the radiation damage problem if we failed to get FIM off the ground. Piers' work on the orientation dependence of damage thresholds and knock-on effects in thin gold films is a separate story. Sadly, Mike Southon died prematurely after a long history of increasingly poor health, while Piers Bowden was killed in an accident shortly after he had joined the staff of the Department of Metallurgy at Cambridge. Mike Wald was a physicist on leave from the Israel Atomic Energy Commission. He was able to contribute some prior experience of radiation damage studies and a great deal of practical good sense. Today, he is Vice-President for Foreign Students at the Fachhochschule, Hamburg.

The group was evacuated after a fire in the laboratory, with no achievements to our credit, and relocated in the old Goldsmiths' Laboratory. I found Mike Wald one day wiring up the reassembled FIM with a strained look on his face. 'Is this wire red or green?' I had never come across colour-blindness before. The microscope was working by the end of the year and we obtained passable FIM images of tungsten and molybdenum at liquid nitrogen temperature. The following year we were joined by two new research students, Brian Ralph and Srinivasa Ranganathan. Brian was a workaholic with a driving ambition to produce 'relevant results'. His task was to examine the possibility of identifying atomic species in alloys, a task which he in part delegated to his wife Anne, who spent many hours counting spots on Brian's FIM images. Ranganathan ('Rangu') was an outstanding Indian student whose mind resembled a box of fireworks — tossing him an idea was like tossing a match into the box. He talked faster than Danny Kaye, and it took all of the rest of us together to keep up with him.

We built a second microscope in no time flat; this time to operate at solid nitrogen temperature (with improved resolution) and to incorporate pulsed field evaporation (so that we could follow the progress of evaporation in the image). Mike Southon constructed our first ultra-high vacuum system so that he could make reliable measurements of field ion currents. He also revised Gomer's calculations of the gas supply to a field-ion specimen tip and derived the first correct analysis of the gas trajectories near the tip. We acquired a polonium α-particle source from Harwell which Mike Wald mounted in our original microscope. We were distinctly nervous, since we had read up on the horrendous medical effects of polonium ingestion, as well as on the self-sputtering propensities of this metal. Within a month we detected contamination in the rotary pump moisture trap and promptly scrapped the whole unit, but not before Mike had imaged damage events in molybdenum which I am still convinced were not an artefact. This was our first and last attempt to study radiation damage directly by FIM.

Our brain-storming sessions took place almost daily in the 'octagonal room' — once the centre of activity of Jack Nutting's electron microscope group. It was here that Mike Southon patiently went over his analysis of field-ionisation for us and Brian Ralph struggled to correlate his messy alloy images with chemical composition and short-range order theories. It was also here that Rangu explained the remarkable degree of atomic order in the grain boundaries which we had fortuitously imaged, and first fell in love with Coxeter's 'compound tessellations'. It was also Rangu who analysed the contrast to be expected from dislocations in the field-ion image.

By 1962 it was obvious that field ion microscopy was going to prove expensive. Ultra-high vacuum was required to prevent field etching of the less refractory metals. Optimum imaging temperatures were for the most part in the range 15 to 25 °K — necessitating good cryogenic control. Image intensification was essential if neon and argon were to be used as imaging gases for iron and steel. Finally, it was now obvious that the identification of chemical species could not be achieved by analysis of image contrast, but would require sensitive mass-spectrometry. This was first achieved by Muller in late 1962, when he developed the 'atom probe', but by the summer of that year I had already decided that it was time to move on.

When I left Cambridge in 1963 Alan Cottrell asked if I had 'lost confidence in field ion microscopy'. It was an embarrassing question. On the one hand, the days of easy research returns for little intellectual effort were clearly over. On the other, the atom probe held out the unique promise of chemical analysis on the atomic scale. Ultimately, it was my confidence in the group, and especially in Mike Southon's good sense, Brian Ralph's drive and Rangu's intelligence, which determined my answer. Cottrell had had the vision to invest in field ion microscopy well before the technique was generally accepted. He later followed this, after I left Cambridge, with massive support for the development of the atom probe by Mike Southon and his colleagues. More personally, he gave me complete freedom to develop research responsibilities at a very early stage in my career and provided wise guidance that has served me well ever since. I learnt how to interpret his gentle hints: 'I don't think that can be quite right' usually referred to a totally fallacious argument, while 'not quite ready for publication' was the mildest of comments on the worst paper I have ever written!

PAPER 9: Creep and Creep Fracture

Dr J. H. Gittus: I will just make a couple of comments. In the late 1950s, I found myself looking after a team of people in the MOND Nickel Laboratories in Wiggin Street developing Nimonic alloys, and we were given the task of trying to increase by 50 °C the temperature at which these alloys were strong enough to be used for turbine blading. We approached this by increasing the amount of gamma prime in the alloys, putting up the amount of gamma prime-forming constituents (titanium and aluminium), and we found a problem which was that, although the creep rate had gone down enormously, they had no creep rupture *life*. They wouldn't last and, of course, blades that don't last are no good to anybody! Microscopy showed that what was happening was that cracks were forming in the grain boundaries and just about that time we saw in the literature a 1957 paper on 'Mechanisms of Grain Boundary Cracking' written by Sir Alan (then Professor) Cottrell. He said there was a critical characteristic length, d: a *sliding* length in the grain boundary. We saw we could reduce that if we had grain boundary precipitates. And because alloy development in those days (I am sure it is quite different now) was just like cookery, we had already made lots of alloys which had precipitates of various sorts in grain boundaries and so we knew how to go about that. We reasoned that if we could make an alloy, in which we had restricted the sliding length in the boundaries by putting a suitable precipitate there, then perhaps that would delay the cracking process and give us a more durable material. Well, that turned out to be the case. Indeed, by good luck the very process that we used for producing the precipitate also gave us a rather soft zone on each side of the grain boundary and it looked as though the strains that would otherwise have been accommodated by grain boundary sliding, were being accommodated by deformation in these thin bands instead. There was a kind of composite material which had been constructed and really what I wanted to say was that the impetus for that, the whole of it really, came from Alan's very elegant and simple explanation of grain boundary cracking. I could go on but I won't take any more time.

Professor G. W. Greenwood: I think those are interesting and relevant comments, highlighting something I could have mentioned more emphatically in my contribution. If you want a ductile material at either low or high temperatures, then a refining of grain size is most important in achieving this. I think many people concerned with power plant materials will agree that the biggest advance in the successful use and in repair of chromium–molybdenum–vanadium steel components has resulted from the control of austenite grain size in welding. Such developments have made a vast difference in practical applications and there are many parallels from the kind of work that you have indicated where different microstructures are produced and subtle effects arise in regions adjacent to the grain boundaries. The length of the sliding remains important, but its influence on cavity nucleation is still a rather grey area and its significance in triple point cracking is not fully resolved.

Professor D. M. R. Taplin: I greatly appreciate this review by Professor Greenwood, covering the many contributions of Sir Alan to 'creep and creep fracture'. Our work in this area on micromechanisms and metallography was very greatly influenced by what I consider to be Sir Alan's most enduring book *The Mechanical Properties of Matter* which I purchased soon after arriving in Australia in 1965. The reason for this particular assertion (against the more widely accepted view regarding the books on 'dislocations and plastic flow in crystals' and on 'theoretical structural metallurgy' is because it united the 'Micromechanistic/physical metallurgy' view of materials with the 'solid mechanics/macro' viewpoint with a unique and forward-looking approach covering for example fluid mechanics and dynamics as well as plasticity and fracture. This was a critical step forward in our thinking in education and research in materials science and engineering. This approach was a particular inspiration for those of us who have worked on materials in Departments of Engineering. I believe that this approach also provided a key contribution to the new synthesis approach of Professor Mike Ashby a decade later continuing a major strand in our thinking in materials. I had an image at that time of Mike Ashby having this book at his bedside for inspiration — (I have never asked Mike whether this is in fact so!).

Sir Alan's contribution to the development of the theory of fracture and to the International Congress on Fracture (founded by Professor Yokobori in 1965) is outstanding and has been recognised by his election by acclamation as an Honorary Fellow of ICF, along with three of his students (John Knott, Mike Ashby and Bruce Bilby), and indeed, one of his 'grand-students' (Rob Ritchie). My own feeling is that it is in the area of fracture, failure analysis and safe design that Sir Alan will be most remembered in the next millenium — Would Professor Greenwood please comment?

Professor G. W. Greenwood: Looking to the future, I feel certain that Sir Alan's contributions will long be remembered and those having direct applicability to the areas you mention of fracture, failure analysis and safe design will surely remain firm cornerstones of the satisfying structure that is now substantially assembled. I note your views of some of Sir Alan's books. All of his books are widely acknowledged to be models of style and clarity, each with a distinctive purpose that I believe makes superfluous any attempt to place them in merit order. It is difficult to envisage a more skilful and valuable interweaving of atomic mechanisms of flow and fracture with the macroscopic behaviour of solids and fluids under stress than that presented in *Mechanical Properties of Matter*. In a different way, the excellent books *Theoretical Structural Metallurgy* and *Dislocations and Plastic Flow in Crystals* respectively brought a stimulating new approach to undergraduate study of metallurgy and provided innovative synthesis of a fast moving research area. One can only admire the range and quality of all his contributions.

PAPER 10: Fibre Reinforcement at the Metallurgy Laboratory at Cambridge 1958–67

Professor J. D. Birchall: If I apply my 'retrospectroscope' it's quite interesting that it's not many years ago that all my group consisted of chemists — I mean, proper chemists — who can tell from smelling the cork whether methyl is on the alpha position. Now it's filled with materials scientists and some metallurgists. The first metallurgist to arrive in my group told the second, what a bunch of absolute idiots we were because we were the sort of chaps who thought a dislocation meant changing laboratories! Three materials scientists, I think, have influenced my group so that it now is largely a group of materials scientists rather than chemists. One is Griffiths, whose equation we have squeezed the last out of. Cottrell is the second and Kelly is the third who has really influenced us. Tony talked about composites and it has been said that composites are 'old hat' (and indeed old hats are composites) and he's taught us how many new tricks there are in composites. I suppose there are three things necessary to make a successful advanced composite. First is a fibre — and that has been one of my contributions, to invent a fibre. Then you need to understand the properties of the matrix into which you are going to put the fibre (e.g. a metal) and I think that has been Sir Alan's contribution. Then you need to understand the properties of the combination and Tony Kelly has done a lot in that area. I decided to make a fibre as a chemist because I couldn't lose. All the materials scientists wanted to put fibres in metals (because the dislocations were too mobile) and into ceramics, because they weren't mobile enough, so I couldn't lose whatever I did. It is interesting too that the highest growth rate in advanced composites right now (I can tell this because I know how much fibre is being used in the world) is with metal matrix composites and I think that must owe a lot to Sir Alan's contribution to understanding how metals behave. One point struck me from this meeting today. In looking at nuclear metallurgy I spotted that NDT was needed to detect 6 mm flaws. If we are going to use ceramics in advanced heat engines we shall need to detect flaws of 6 microns, so there is a lot of work left to do. We need new fibres for ceramic composites and new process technologies for putting them in a ceramic matrix. I am glad to say that all this needs chemical tricks to accomplish.

Dr P. W. R. Beaumont: In the summer of 1963, in a paper to the Royal Society,* Sir Alan Cottrell carefully defined the requirements of a solid having a high bulk strength and toughness in the presence of cracks: '*... it is necessary to build into the material an alternative mode of failure (to brittle cleavage), weaker in so far as it has a lower crack tip failure stress, but with a much larger crack tip fracture displacement ... The body is then trapped in a microscopically weak but macroscopically strong mode of fracture*'.

A few months later, a series of experiments at the Royal Aircraft Establishment, Farnborough, England, led to the discovery in the following Spring of high strength, high stiffness carbon fibres and their composites. Despite the natural fragility of these fine fibres, their mixture into a brittle polyester resin produced a strong fibrous composite with a work of fracture two orders of magnitude above that of the pure matrix alone. As Sir Alan had indicated earlier, crack propagation in a fibrous material can be accompanied by the withdrawal of long fibre ends from their holes over distances that depend on the restraining tangential force at the interface. What happens, of course, is that the matrix cracks once formed are rendered innocuous by the intact strong fibres which bridge them and still transmit stress across. Such bridging of a crack by fibres leads to a crack tip stress-shielding effect. Furthermore, a crack wake microprocess, e.g. Sir Alan's pull-out mechanism, is still able to interact with mechanisms that operate in the crack tip zone because the fibres still transmit a stress. The result is an increase in the size of the tip process zone and crack bridging zone to provide synergistic (multiplicative) toughening of the composite. This interaction between mechanisms is only now being recognised and exploited in the development of organic, metallic and new toughened ceramic composite systems, 25 years after the original paper of Sir Alan.

Dr J. A. Charles: On joining the University of Cambridge Department of Metallurgy in 1960 at the invitation of Sir Alan, then Goldsmiths' Professor, I was quickly made aware of the significance that he and other

---

* *A Discussion of New Materials* held at the Royal Society, 6 and 7 June 1963, and published in *Proc. Roy. Soc. Lond.* **A282**, 20 October 1964.

colleagues, such as Tony Kelly, apportioned to the development of composite structures in metallic materials, essentially consisting of strong fibres in a ductile matrix. In my own case my background and interests suggested a study of the production of controlled microstructures in steel, with the late Dr R. L. Cairns, utilising the system of directional martensitic regions in a ferrite matrix.

Good control of the microstructure was possible by making a synthetic pack-rolled composite, one component containing a ferrite-stabilising element such as aluminium and alternating with a high carbon steel. Such materials could be quenched from an austenite (high-carbon)/ferrite (low-carbon) pseudo-equilibrium state to give the desired structure. Some success was also achieved in producing designed ferrite-martensite microstructures by quenching heavily-banded steel plate from a short-time annealing treatment. This latter approach was to be the basis of treatment subsequently used for the production of stiffer dual-phase steel for car bodies, particularly front and rear structural members where the greater benefit of the superior energy-absorbing characteristics could be obtained, justifying the higher cost. In our work it was clear, however, that a large degree of order in the microstructures could only be achieved by synthetic production methods.

Mechanical tests showed that plain-carbon martensite could carry higher stress when protected from brittle failure, by being situated within a more ductile phase. A linear variation of certain properties, over quite large ranges of martensite volume fraction indicated that the broad principles of fibre-strengthening applied to the behaviour of these composites.

This work, like so much else in the Department at that time, owed much to the direct encouragement of Sir Alan.

PAPER 11: Strengths of Grain Boundaries in Pure Metals

Dr E. D. Hondros: Thank you Chairman. Having been closely involved in this sort of work and listening carefully to Sir Alan's paper, my overwhelming feeling is akin to that of the wedding guest in the poem of Coleridge who wakes up the following morning, a holier and a wiser man (or something like that). I have been kicked by the flash of insight demonstrated here into realising that the otherwise simple and unfinished ideas that Donald McLean and I put forward in the quoted paper can be extended to show elegantly why some metals are intrinsically brittle while others are tenacious. Sir Alan has introduced a fresh and original departure — thank you for enlightening us once more.

It should be added that our main concern in the quoted work was to try to understand why grain boundaries in common industrial alloys were tenacious in certain conditions of heat treatment, temperature, and bulk impurity content, while in other conditions of processing or even during operation, the cohesion became inadequate, leading to spectacular examples of intergranular failure. This led us to focus on the balance between the free energy of the pre-existing grain boundary and the free energies of the two surfaces created by fracture. These two types of interface are affected by Gibbsian equilibrium adsorbtion of active impurity species, but for a variety of physico-chemical reasons, the effects on the surface and grain boundary free energies may differ extensively in magnitude. If the surface energy term is reduced significantly by the solute adsorbed at the prior grain boundary, there is potentially a weak grain boundary. In the case mentioned by Sir Alan of carbon strengthening the grain boundary, our view would have been that the differential effects of carbon adsorption are such that the surface energy term remains high and so does the work of fracture. There is, however, insufficient experimental data to test this.

Chairman, on another note, since all speakers today have taken liberally the opportunity to indulge in sentiments on the occasion of this event, let me join in. First, let me say that I am probably the only one here who has at no time been a student of Sir Alan, nor have I at any time been associated as a research colleague — therefore, you may ask (apart from having been invited to attend) what am I doing here? In fact I have a good reason to be here; I have purposely flown over from Amsterdam today to pay homage to a great man, for I have lived long enough on the Continent to realise that there is a responsibility for us to honour scholars in their lifetime (perhaps a rare event in this country). Hence the main reason for my being here is to underline the fact that there exists a vast number of scientists outside these four walls who have been influenced in their life's work and career by the writings of Sir Alan. I speak up in gratitude and on behalf of people outside this meeting.

To indulge further, in common with all my colleagues here, let me say something personal. Many years ago, literally on an Australian shore, I opened a book, having been encouraged by Trevor Broom (present in the audience) to abjure my wasteful habits and to learn something useful. So I opened *Theoretical Structural Metallurgy* by A. H. Cottrell and read the book through, following which I can only say that I had the same feeling as the character in the Keats sonnet who climbed the mountain and gazed at the Pacific, 'silent upon a peak in Darien'. At the risk of mixing literary allusions, I use this rhetoric to describe that rare feeling of revelation. The same impact has been felt I am sure by hundreds of readers of the published works of Sir Alan, who has paved many paths in this fascinating field of materials science.

Let me pay homage to Sir Alan, personally, and on behalf of the many who are not here.

Dr Robert S. Barnes: At this late stage in the proceedings I merely wished to add a personal note to all that has gone before. However, listening to Alan's contribution brought back so clearly the stimulation and excitement he so often generated. A few simple, well expressed thoughts would start the imagination and lead to ideas and experiments which led one further. The concepts which Alan has put forward, often first to some of us here, have lead to exciting and important developments.

I knew Alan when he was still at Birmingham, but it was when he came to Harwell in 1955 that I got to know him really well. We worked very closely together and he had a considerable impact. He joined a team which was extremely young, full of vigour and enthusiasm. He was a mature thirty-four, while few of us were over thirty, yet with generous resources and with considerable responsibility. The work was on the frontiers and it was easy to become an instant expert. There were so many opportunities, whatever one did turned up something new and unexpected. The flair needed was to recognise and go for those things which were to be important, all were interesting and many exciting. This was particularly the case for work on the radiation of materials where the effects of radiation on man and on materials was hardly known and certainly not understood.

To illustrate the atmosphere of the times, and as it has already been referred to, I recall my attempts to understand the effect on internal friction caused by the locking of dislocations by point defects (both vacancies and interstitial atoms) produced by radiation. Certainly, it needed but a few minutes of radiation to lock the dislocations and markedly reduce the internal friction. For instance, well annealed copper could be made to 'ring' like a bell. I remember particularly my first reactor radiation experiments to be performed at liquid nitrogen temperature. I took to the top of the nuclear reactor my well annealed rod of Spec. pure Johnson Matthey Copper in a small Dewar flask containing liquid nitrogen. With a Health Physicist alongside monitoring the radiation level (actually he was behind as it was safer), I lowered the flask, attached to the end of a piece of string, into one of the reactor channels. After a few minutes there was a dull thud from down below. I hauled up my experiment which was still intact, but naturally quite 'active'. This might have been my most spectacular experiment.

Alan arrived with qualities not expected in a University Professor (not in those days anyway), a professionalism, a realism and above all a clear view of what was likely to be of practical importance. The materials which could be used were most restricted and their behaviour in the intense neutron environment was unknown, let alone understood. He single-mindedly focused on those areas which were to be of vital importance to the harnessing of nuclear energy. We all remember his pioneering contributions to the radiation induced creep of Uranium, the storage of energy in Graphite, and the radiation induced brittleness of steel, which was so critical to the continuing well being of reactor pressure vessels, particularly those of the American pressurised water reactors. It is difficult to appreciate that he was only at Harwell for some three years, so many things happened in that short time. Perhaps some of them happened afterwards, but only because he paved the way.

Of course, I have kept in touch with Alan since he left Harwell, during his time at Cambridge, examining his students, in Whitehall while he was at the Ministry of Defence and later as Scientific Advisor to HMG, then back to Cambridge, where I suspect he will be for ever. One could always discuss one's technical problems with Alan and receive clear analysis and insight. I am sure that many of you here have so benefitted and some will be grateful for his helpful advice on their career choices. Here his advice was not only sound, but all was kept completely confidential.

I should like to thank you, Alan, on behalf of all of us, not only for your contribution to Science which has been outlined here today but also for the great impact you have had on its application wherever materials are used. Thank you for the inspiration you have given to us.

Concluding Remarks

Professor R. W. K. Honeycombe

Emeritus Goldsmiths' Professor of Metallurgy, University of Cambridge

Ladies and Gentlemen:

What a splendid day we've had, when a very wide range of subjects has been discussed in all of which Alan Cottrell played a profound part! Many of us here today have known Alan for over forty years, so we have had ample time to appreciate to the full the depth and breadth of his outstanding contributions to the study of materials.

Most of the topics dealt with today were already under scrutiny by Alan in 1950. I well remember arriving from Australia in 1949 to find in the UK an impressive post-war resurgence in the basic study of materials, particularly in Birmingham, Cambridge and Bristol. Alan was already one of the established leaders of this renaissance which included other famous names such as Mott, Orowan, Frank and Nabarro. In particular I remember writing back to my former Professor in Melbourne — Neill Greenwood — with great enthusiasm about the new theory of the sharp yield point, a phenomenon which had defied satisfactory explanation until the elegant Cottrell theory emanated from Birmingham.

Dislocations had been predicted by Polanyi, Orowan and G. I. Taylor in 1934 but World War II put paid to the development of the original concept. Several of us here today were working in Egan Orowan's group in the Cavendish Laboratory in 1948 about 50 yards from where I am now standing. Orowan used to tell us that he developed his original ideas in his doctoral thesis in Hungary in 1934. Being somewhat sceptical, members of the group felt a little scientific investigation was required so one of our members was given the job of locating a copy of the elusive thesis. He succeeded and there to our collective surprise was the sound evidence! But the definitive work on this subject was Alan Cottrell's *Dislocations and Plastic Flow in Crystals* published in 1953, which had a tremendous impact on the subject both in teaching and research.

Another vivid memory of the late forties and early fifties was the large number of meetings on these matters — very interdisciplinary — physicists, metallurgists, chemists, not only from the UK but also France, Germany and USA as the interest grew, took part. The enthusiasm and impact of these meetings were memorable and in all Alan played a leading part. A lot of water has flowed down the Cam since those heady days and times have changed. It is now rarely possible to experience so strongly the feeling that science is actually in the making during a scientific meeting.

Much of this classical work in our subject has now borne fruit, and concepts originally regarded solely of academic interest have been generally accepted, moreover many have merged into the technology of the subject sometimes in surprisingly direct ways. Anyone who suggested in 1950 that turbine blades for jet engines would be made from single crystals would have been considered to be of the lunatic fringe. Yet today we have a high technology industry which has achieved this objective with great precision and in this way gained substantial technical advantage.

Alan's long and wide experience in government circles has given him an excellent opportunity to encourage the introduction of modern science and technology into various national activities. Inevitably, such efforts are not exposed to scrutiny as is laboratory-based work, but they are particularly important to ensure that the country derives the maximum benefit from the large investment in scientific research. But one has the feeling that Alan is happiest when worrying away at some scientific problem until he reaches a satisfactory solution. He still possesses this capacity to elucidate as he has shown in cameo this afternoon. Moreover in his retirement he has again illustrated his strength as a writer of text books by producing a new book on electrons in metals. I am sure that like all its predecessors it will be influential and widely used.

To reach seventy is a landmark in anyone's life, but to reach it with an active and original mind is to be greatly admired, and we trust that Alan will continue in this vein for a long time to come. Perhaps I should remind him that Guiseppe Verdi composed his greatest opera *Otello* in his 74th year followed by *Falstaff* in his 80th year. So perhaps Alan has a tour-de-force tucked up his sleeve for the future.

Be that as it may, I think that today's proceedings have reminded us of many facets of Alan's career in an admirable way which has done full justice to his scientific achievements. May I now conclude the meeting by warmly thanking all the contributors for their talks which have provided a very fitting tribute to Alan in the year of his 70th birthday. Finally, we extend to Alan and Jean our very best wishes on this special day, and express the hope that they will spend many pleasant years here in Cambridge.

Publications of Professor Sir Alan Cottrell

(to June 1989)

1940 — The Recovery of Metals from the Effects of Cold Work, *J. Inst. Metals* **66**, 163.

1941 — The Arc Welding of High Tensile Alloy Steels (jointly with E. C. Rollason, K. Winterton and P. D. Crowther), *Trans. Inst. Welding* **4**, 3.

1944 — Researches in Alloy Steel Welding (jointly with D. Hanson, K. Winterton, P. D. Crowther and J. A. Wheeler), *Trans. Inst. Welding* **7**, 44.

— Tensile Properties of Unstable Austenite. *J. Iron Steel Inst.* **1**, 93.

1946 — What is a Metal?, *J. Birm. Met. Soc.* **26**, 3.

1947 — Thermal Disorder in Metals and Alloys, *Metal Treatment*.

— Andrade's Creep Law and the Flow of Zinc Crystals (jointly with V. Aytekin), *Nature* **160**, 328.

1948 — Theoretical Structural Metallurgy. Edward Arnold, London.

— Effect of Solute Atoms on the Behaviour of Dislocations. Physical Society Bristol Conference Report, 30.

— Thermal Hardening of Cadmium Crystals (jointly with D. F. Gibbons), *Nature* **162**, 488.

1949 — Dislocation Theory of Yielding and Strain Ageing of Iron (jointly with B. A. Bilby), *Proc. Phys. Soc.* **A62**, 49.

— Distribution of Solute Atoms round a Slow Dislocation (jointly with M. A. Jaswon), *Proc. Roy. Soc.* **A199**, 104.

— Change of Electrical Resistance during the Strain Ageing of Iron (jointly with A. T. Churchman), *J. Iron Steel Inst.* **9**, July, 271.

— Theory of Dislocations, in *Progress in Metal Physics — I*, Butterworths, London.

1950 — Yield Points in Zinc Crystals (jointly with H. L. Wain), *Proc. Phys. Soc.* **B63**, 339.

— Surface Effects and the Plasticity of Zinc Crystals (jointly with S. Harper), *Proc. Phys. Soc.* **B63**, 331.

— The Flow of Zinc under Constant Stress (jointly with V. Aytekin), *J. Inst. Metals* **77**, 389.

1951 — The Yield Point in Single Crystals and Polycrystalline Metals. Report of Pittsburgh Conference on Plastic Deformation of Crystals.

— Yield Phenomena and Twinning in Alpha Iron (jointly with A. T. Churchman), *Nature* **167**, 943.

— A Mechanism for the Growth of Deformation Twins (jointly with B. A. Bilby), *Phil. Mag.* **42**, 573.

1952 — The Formation of Immobile Dislocations during Slip, *Phil. Mag.* **43**, 645.

— The Yield Point in Single Crystals and Polycrystalline Metals. Report of Ninth Solvay Conference.

— The Time Laws of Creep, *J. Mech. Phys. Solids* **1**, 269.

— The Effect of Quench Ageing on Strain Ageing in Iron (jointly with G. M. Leak), *J. Iron Steel Inst.* **172**, 301.

1953 — Dislocations and Plastic Flow in Crystals. Oxford.

— Work-Softening in Aluminium Crystals (jointly with R. J. Stokes), *Acta Met.* **2**, 341.

— Theory of Dislocations, in *Progress in Metal Physics — IV*. Butterworths, London.

— Metals, Alloys, Slags and Fluxes. The Faraday Society.

— Yield Point in Brass Crystals (jointly with G. W. Ardley), *Proc. Roy. Soc.* **A219**, 328.

— A Note on the Portevin–Le Chatelier Effect, *Phil. Mag.* **44**, 829.

— Electrical Interaction of Dislocations and Solute Atoms (jointly with F. R. N. Nabarro and S. C. Hunter), *Phil. Mag.* **44**, 1064.

1954 — Work-Hardening in Stretched and Twisted Aluminium Crystals (jointly with H. W. Paxton), *Acta Met.* **2**, 3.

— Interactions of Dislocations and Solute Atoms. ASM Symposium on Relation of Properties to Microstructure.

1955 — Effects of Temperature on the Plastic Properties of Aluminium Crystals (jointly with R. J. Stokes), *Proc. Roy. Soc.* **A233**, 17.

— Annealing of Point Defects (jointly with W. M. Lomer), *Phil. Mag.* **46**, 711.

— Effect of Temperature on the Flow Stress of Work Hardened Copper Crystals (jointly with M. A. Adams), *Phil. Mag.* **46**, 1187.

— A Study of Primary and Conjugate Slip in Alpha Brass (jointly with G. R. Piercy and R. W. Cahn), *Acta Met.* **3**, 331.

— Quench Hardening of Aluminium Crystals (jointly with R. Maddin), *Phil. Mag.* **46**, 735.

1956 — Slip in Crystals of the Caesium Chloride Type (jointly with W. A. Rachinger), *Acta Met.* **4**, 109.

— Creep of Alpha Uranium during Irradiation with Neutrons

(jointly with A. C. Roberts), *Phil. Mag.* **1**, 711.

— Dislocations in Crystals. IUTAM Madrid Colloquium on Deformation and Flow of Solids.

— Creep and Ageing Effects in Solid Solutions. NPL Symposium on Creep and Fracture of Metals at High Temperatures.

— Dislocations and the Mechanical Properties of Crystals, *Nature* **178**, 1090.

— The Effects of Irradiation on the Physical Properties of Solids, *Brit. J. Appl. Phys.* No. 5, 43.

— Effects of Neutron Irradiation on Metals and Alloys, *Inst. Met. Metall. Revs.* **1**, 479.

1957 — Point Defects and the Mechanical Properties of Metals at Low Temperatures. The Institute of Metals, Monograph 23.

— Extrusion and Intrusion by Cyclic Slip in Copper (jointly with D. Hull), *Proc. Roy. Soc.* **A242**, 211.

— The Solution of Inert Gas Atoms in Metals (jointly with D. Rimmer), *Phil. Mag.* **2**, 1345.

— Hardening and Embrittlement of Steel by Irradiation with Neutrons (jointly with A. T. Churchman and I. Mogford), *Phil. Mag.* **2**, 1271.

— The Deformation of Solids at High Rates of Strain. Conference on High Rates of Strain, *Inst. Mech. Eng.*

— The Intersection of Gliding Screw Dislocations. Lake Placid Conference on Dislocations and Mechanical Properties of Crystals.

1958 — Theory of Brittle Fracture in Steel and Similar Metals, *Trans. AIME*, April, 192.

— The Observation of Vacancy Sources in Metals (jointly with R. S. Barnes and G. B. Redding), *Phil. Mag.* **3**, 97.

1959 — Effects of Nuclear Radiation on Engineering Materials. Institution of Mechanical Engineers, Hawksley Lecture.

1960 — The Strength of Solids, *Proc. Roy. Inst.* **38**, 346.

— The Metallic State, *Contemp. Phys.* **1**, 417.

— Scientists, Solo or Concerted? *The Listener* **64**, 411.

1961 — Theoretical Aspects of Radiation Damage and Brittle Fracture in Steel Pressure Vessels. Iron and Steel Institute Special Report No. 69.

— Intercrystalline Creep Fractures. Iron and Steel Institute Special Report No. 70.

1962 — Stored Energy in the Graphite of Power-Producing Reactors (jointly with J. C. Bell, H. Bridge, G. B. Greenough, W. N. Reynolds and J. H. W. Simmons), *Phil. Trans. Roy. Soc.* **A254**, 361.

1962 — Fracture, in *Properties of Reactor Materials and The Effects of Radiation Damage*, Butterworths, London.
— Field-Ion Microscopy, *J. Inst. Met.* **90**, 449.

1963 — Fracture (Bakerian Lecture), *Proc. Roy. Soc.* **A276**, 1.
— Mechanisms of Fracture. Tewksbury Lecture, University of Melbourne.
— Discontinuous Yielding. NPL Conference on Relations between Structure and Strength, HMSO.
— The Spread of Plastic Yield from a Notch (jointly with B. A. Bilby and K. H. Swinden), *Proc. Roy. Soc.* **A272**, 304.
— Notch Brittleness in Mild Steel (jointly with J. F. Knott), *J. Iron and Steel Inst.* **201**, 249.

1964 — The Mechanical Properties of Matter. John Wiley, New York.
— Theory of Crystal Dislocations. Gordon and Breach, London.
— Strong Solids. *Proc. Roy. Soc.* **A282**, 2.
— Plastic Yielding from Sharp Notches (jointly with B. A. Bilby, E. Smith and K. H. Swinden), *Proc. Roy. Soc.* **A279**, 1.
— The Liquid State, *The British Foundryman*, 259.

1965 — Mechanisms of Fracture in Large Structures, *Proc. Roy. Soc.* **A285**, 10.
— Mechanisms of Fracture, in *Proc. First Tewksbury Symposium*. Butterworth, London.
— Elastic Failure at Notches in Silicon Steel (jointly with J. R. Griffiths), *J. Mech. Phys. Solids* **13**, 135.
— The Ultimate Mechanical Properties of Solids, in *Metallurgical Achievements*, Pergamon, Oxford.

1966 — Le Microscope à Emission d'Ions, *Revue de Metallurgie* **63**, 237.
— The Design of Strong Materials (jointly with A. Kelly), *Endeavour* **25**, 27.
— Science and Economic Growth in the United Kingdom. *Adv. Sci.*, Brit. Assn.

1967 — Ductile and Brittle Crystals (jointly with W. R. Tyson and A. Kelly), *Phil. Mag.* **15**, 567.
— An Introduction to Metallurgy. Edward Arnold, London.
— Notch Effects on Slip, Twinning and Fracture in Silicon Steel (jointly with J. R. Griffiths), *J. Mech. Phys. Sol.* **15**, 125.
— The Nature of Metals, *Scient. Amer.* **217**, 90.
— Strength of Materials, *Contemp. Phys.* **8**, 1.

1969 — The Structure of a Crack, in *Physics of Strength and Plasticity*, M.I.T. Press, Cambridge, MA.
— Science and Technology, *J. Roy. Soc. Arts* **117**, 173.
— Simple Physics in a Complex World, *Phys. Bull.* **20**, 362.
— Technological Thresholds, in *The Process of Technological Innovation*, Nat. Acad. Eng., Nat. Acad. Sci., Washington.

1970 — Competition and Cooperation in Technological Innovation, in *International Aspects of Technological Innovation*, UNESCO.

1972 — Edward Neville da Costa Andrade, *Biog. Mem. Roy. Soc.* **18**, 1.

1973 — Materials and Energy, *Met. Trans. AIME* **4**, 405.

— World Energy Resources and their Challenge to Science. Berkeley Conference. The Metals Society.

— The Age of Scarcity. Institute of Mining and Metallurgy, Minerals and Environment Symposium.

1974 — Fracture of Steel Pressure Vessels. Memorandum to Select Committee on Science and Technology, First Report 1973–4, HMSO.

— The Age of Scarcity, *Trans. Inst. Min. Met.* **83**, 25.

— Long-Term Trends in Materials Resources, *Delta Review*, 1974–5.

1975 — Portrait of Nature. Collins, London.

— Fracture, in *The Physics of Metals — 2 (Defects)*. Edited by P. B. Hirsch, Cambridge University Press, Cambridge.

— World Energy Resources and their Challenge to Science, in *Physical Metallurgy of Reactor Fuel Elements*, The Metals Society, London.

1976 — The Voice of the Engineer in Public Policy. 21st Graham Clark Lecture, Council of Engineering Institutions.

— Grasping the Nettles in British Industrial Policy. Royal Society of Arts, March 1976.

— The Contribution of Science to the Economy: A United Kingdom Point of View, *Interdisciplinary Sci. Rev.* **1**, 12.

— Technical Lessons of Flixborough: A Metallurgical Examination of the 8 Inch Line (jointly with P. R. Swann), *The Chemical Engineer*, April.

— The Task for the Educator: Rosenhain Centenary Conference, *Phil. Trans. Roy. Soc.* **A282**, 467.

— The Rise and Fall of Science Policy, *New Scientists* **72**, 80.

— The Science of Complex Matter, *J. Sheff. Univ. Met. Soc.* **15**, 33.

1977 — Engineering as a Source of Scientific Ideas, *Interdisciplinary Sci. Rev.* **2**, 94.

— Emergent Properties of Complex Systems, in *The Encyclopaedia of Ignorance* (Edited by Ronald Duncan and Miranda Weston-Smith), Pergamon Press, Oxford.

1978 — Environmental Economics. Edward Arnold, London.

— Ten Years Tinkering with the Engines of Industry, in *What's Wrong with Britain*, Edited by Patrick Hutber, Sphere Books Ltd, London.

1979 — The Undermining of British Industry. The 25th Fawley Foundation Lecture, University of Southampton.

— The Natural Philosophy of Engines, *Contemp. Phys.* **20**, 1.
— Science is Objective, in *Lying Truths* (Edited by Ronald Duncan and Miranda Weston-Smith), Pergamon Press, Oxford.
1980 — Science Policy in Britain, *Proc. Roy. Inst.* **52**, 179.
— Which Nuclear Reactors for Britain? *Coal and Energy Q.* **25**, 3.
— Introductory Address, Future Metals Strategy. The Metals Society, London.
1981 — Research, Risk and Reward, *Chemistry and Industry* **42**.
— Britain at Malthus Two. The Foundation Lecture, Birkbeck College.
— Ulick Richardson Evans (jointly with J. E. O. Mayne), *Biographical Memoirs R.S.* **27**, 235.
— Annealing a Nuclear Reactor: an Adventure in Solid-State Engineering, *Nucl. Mat.* **100**, 64.
1983 — Brittle Fracture from Pile-ups in Polycrystalline Iron, in *Yield, Flow and Fracture of Polycrystals.* (Ed. T. N. Baker) Applied Science Publishers,
— Engineering as a Pure Science, *The Cambridge Review* **47**, 28 February.
— Materials Research for Fusion Reactors. Report to the International Energy Agency (OEC).
1984 — The Sorcerer's Apprentice, in *Science and Politics*, The Herbert Spencer Lecture, 1982 (Edited by Vernon Bogdanor), Oxford.
— The Physical World and Human Experience. The South Place Ethical Society.
— Geoffrey Vincent Raynor, *Biog. Mem. Roy. Soc.* **30**, 545.
1985 — Closing Address, in *Dislocations and Properties of Real Materials.* The Institute of Metals, London.
— Public Attitudes to Energy. A lecture, published by the Central Electricity Generating Board.
1986 — Thermodynamics as Engineering Science, in *Teaching Thermodynamics* (Edited by Jeffery Lewins), Plenum Press, New York.
— Government Research and Development Reorganisation in the U.K. in 1972, *Science and Public Policy* **13**, 55.
— Scaling Laws and Structural Inhomogeneities in Solids, *South African J. Phys.* **9**, No. 2, 44.
— The Case for Nuclear Power in Britain, *The Observer*, 28 September.
1987 — Safe Energy, in The University of Wales Science and Technology Review, Number 2, Autumn.
— Warm Prestressing and Fracture in Structural Steel, *Engng. Fracture Mech.* **28**, 481.
1988 — Introduction to the Modern Theory of Metals. The Institute of Metals, London.
— Heading towards a Nuclear Dead-End, *Financial Times*, 14 December.

For Product Safety Concerns and Information please contact our EU representative GPSR@taylorandfrancis.com Taylor & Francis Verlag GmbH, Kaufingerstraße 24, 80331 München, Germany

Printed and bound by CPI Group (UK) Ltd, Croydon, CR0 4YY

23/10/2025

01983295-0016